Elements of Superintegrable Systems

Mathematics and Its Applications

B. A. Kupershmidt
The University of Tennessee Space Institute,
Tullahoma, U.S.A.

Elements of Superintegrable Systems

Basic Techniques and Results

D. Reidel Publishing Company

A MEMBER OF THE KLUWER ACADEMIC PUBLISHERS GROUP

Dordrecht / Boston / Lancaster / Tokyo

Library of Congress Cataloging in Publication Data

CIP

Kupershmidt, Boris A., 1946-
Elements of superintegrable systems.

(Mathematics and its applications)
Bibliography: p.
Includes index.
1. Differential equations, Partial. 2. Hamiltonian systems.
3. Lie algebras. I. Title. II. Title: Superintegrable systems. III. Series.
QA377.K85 1987 515.3'53 86-33930
ISBN-13: 978-94-010-8190-0 e-ISBN-13: 978-94-009-3799-4
DOI: 10.1007/978-94-009-3799-4

Published by D. Reidel Publishing Company,
P.O. Box 17, 3300 AA Dordrecht, Holland.

Sold and distributed in the U.S.A. and Canada
by Kluwer Academic Publishers,
101 Philip Drive, Assinippi Park, Norwell, MA 02061, U.S.A.

In all other countries, sold and distributed
by Kluwer Academic Publishers Group,
P.O. Box 322, 3300 AH Dordrecht, Holland.

To Yu. I. Manin from a respectful distance

Series Editor's Preface

Approach your problems from the right end and begin with the answers. Then one day, perhaps you will find the final question.

'The Hermit Clad in Crane Feathers' in R. van Gulik's *The Chinese Maze Murders.*

It isn't that they can't see the solution. It is that they can't see the problem.

G.K. Chesterton. *The Scandal of Father Brown* 'The point of a Pin'.

Growing specialization and diversification have brought a host of monographs and textbooks on increasingly specialized topics. However, the "tree" of knowledge of mathematics and related fields does not grow only by putting forth new branches. It also happens, quite often in fact, that branches which were thought to be completely disparate are suddenly seen to be related.

Further, the kind and level of sophistication of mathematics applied in various sciences has changed drastically in recent years: measure theory is used (non-trivially) in regional and theoretical economics; algebraic geometry interacts with physics; the Minkowsky lemma, coding theory and the structure of water meet one another in packing and covering theory; quantum fields, crystal defects and mathematical programming profit from homotopy theory; Lie algebras are relevant to filtering; and prediction and electrical engineering can use Stein spaces. And in addition to this there are such new emerging subdisciplines as "experimental mathematics", "CFD", "completely integrable systems", "chaos, synergetics and large-scale order", which are almost impossible to fit into the existing classification schemes. They draw upon widely different sections of mathematics. This programme, Mathematics and Its Applications, is devoted to new emerging (sub)disciplines and to such (new) interrelations as exempla gratia:

- a central concept which plays an important role in several different mathematical and/or scientific specialized areas;
- new applications of the results and ideas from one area of scientific endeavour into another;
- influences which the results, problems and concepts of one field of enquiry have and have had on the development of another.

The Mathematics and Its Applications programme tries to make available a careful selection of books which fit the philosophy outlined above. With such books, which are stimulating rather than definitive, intriguing rather than encyclopaedic, we hope to contribute something towards better communication among the practitioners in diversified fields.

A prime example of a field where a most unusual number of disciplines come together is "integrable system theory". Modestly classified under 58F07 somewhere in global analysis it involves in essential ways: (i) combinatorics in the form of a number of fascinating identities known as Rogers-Ramanujan and Macdonald identities; (ii) algebraic geometry via theta functions, abelian varieties and flexes; (iii) Lie algebras, via both symmetry considerations and orbits of the co-adjoint representation; (iv) infinite dimensional Lie groups in the form of loop groups; (v) functions of a complex variable via the Riemann-Hilbert boundary value problem and the dressing method; (vi) ordinary differential equations via deformations and Painlevé transcendants; (vii) partial differential equations via overdetermined systems and the $\bar{\partial}$-Neumann problem, and also pseudo-differential operators; (viii) special functions via (again) τ-functions but also Schur polynomials and hence the representation theory of the symmetric and general linear groups; (ix) integral equations via the

Gelfand-Levitan-Marcenko equation of inverse scattering and the singular integral equations at the basis of the direct linearization methods; (x) calculus of variations via the formal variational calculus approach initiated by Gelfand and a host of coworkers; (xi) global analysis because that is what it is all about; (xii) quantum mechanics because of the existence, usually, - or quite possibly always, as far as we know - of quantized versions of systems and because a number of important model equations are integrable; (xiii) fluid mechanics, because an important number of model equations in this field are integrable; (xiv) dynamics of rigid bodies, again because a number of important examples are integrable, e.g. the Euler, Lagrange and Kowalevskaya top; (xv) statistical physics because of the link with various exactly solvable models such as the eight-vertex model and the link with the Yang-Baxter equations. This is far from a complete summing up and I am aware of various links, quite important ones, with still other areas of mathematics. There are, no doubt, others I know nothing about.

The above paragraph concerns just "non-super" integrable systems. However it has, become clear that substantial parts of mathematics have super-counterparts where both ordinary commutative functions occur and odd functions which anticommute: $\mathbb{Z}/(2)$ graded versions of algebra, analysis and geometry, of which the even part constitutes, roughly speaking, ordinary algebra, analysis, geometry, ... The odd parts correspond to fermions in physics and superthinking is assuming epidemic proportions in physics.

Thus we have super-Lie-algebras, superspace, supermanifolds, supergroups, ..., and super-integrable systems, the topic of this book.

Yet it is still a book which can be used as a starting volume for someone who wants to get a real grip on the manifold world of integrable systems, though of course it cannot describe in full detail all the varied aspects listed above.

Summing up I can say that it is a book I am really happy to be able to present to the mathematical community at large via this series.

The unreasonable effectiveness of mathematics in science ...

Eugene Wigner

Well, if you know of a better 'ole, go to it.

Bruce Bairnsfather

What is now proved was once only imagined.

William Blake

As long as algebra and geometry proceeded along separate paths, their advance was slow and their applications limited.

But when these sciences joined company they drew from each other fresh vitality and thenceforward marched on at a rapid pace towards perfection.

Joseph Louis Lagrange.

Bussum, December 1986

Michiel Hazewinkel

CONTENTS

PREFACE

The last two decades have seen the creation of a new area of mathematical physics: the theory of infinite-dimensional integrable systems, both continuous and discrete. The field is still developing swiftly (detailed accounts of various developments can be found in [Man; Mc; No; La; A-S; Ne]), and new territories are being added with ever increasing speed. The two latest additions are the Yang-Baxter (YB) equations and superintegrable systems. The current state of the YB equations can be pieced together from the beautiful recent papers [D; J] which study the quantization of the classical YB equations; the interested reader can obtain more complete information by tracing out the references in these two articles. Our subject in this book is the theory of continuous superintegrable systems.

The text is intended to be a fully detailed introduction to the basic techniques, ideas, and results of the theory. Each of the two Chapters is devoted to one of the two principal themes.

Chapter I, on super Lax equations, is designed both to improve and to parallel the ideal, if not historical, logical sequence in the development of the usual, even (i.e., not super) theory; so that the reader can efficiently learn the latter if he is unfamiliar with it, or has encountered difficulties in following the existing literature. For such readers I recount the highlights of the even theory in Section 1, including what I think is by far the deepest and most fundamental result to date: Wilson's definition and classification of matrix Lax equations (difficult to guess, easy to state, difficult to prove, everything else follows from it). For the reader interested primarily in learning the Lagrangian and Hamiltonian formalisms in the presence of odd (i.e., super) variables, Sections 2 and 3 provide a self-contained manual. In Section 4 we complete the development of the basic tools indispensable for *any* theory dealing with integrable systems, be they super or not. The super Lax equations are introduced and classified in Section 5. The tools honed in Sections 2 through 4 are employed in Section 6 to derive the Hamiltonian structure of the super Lax equations, to prove the involutivity of the infinite number of conservation laws constructed in Section 5, and then to tackle the notoriously difficult problem of analyzing nontriviality of these conservation laws.

Chapter II deals with various connections between Lie algebras and integrable systems, both super and even. The central idea is to use generalized two-cocycles on differential Lie algebras to construct infinite hierarchies of commuting integrable bi-Hamiltonian systems. The exact recipe is extracted in Section 7 by a careful analysis of the bi-Hamiltonian structure of the Korteweg-de Vries (KdV) equation. This recipe is then utilized to bring forth three types of integrable systems associated to a large class of Lie algebras (including semisimple ones). To prove that the constructed systems are indeed integrable, we develop two subjects important in their own right: Identities satisfied by special functions on the dual space to a Lie algebra endowed with an invariant nondegenerate metric (Section 8), and the theory of bi-superHamiltonian systems (Section 9). As a simple application, a new infinite superintegrable hierarchy of super Harry Dym equations is derived in Section 9.

A few words follow about the format of the book. Each section (except section 1) is preceded by a short summary of new notions and results. The part called "Sources" provides an account of the historical developments of the theory on a section-by-section basis, together with additional references. I have tried to make the book as self-contained as possible by providing plenty of motivation on the way. The knowledge of the even theory of integrable systems is neither assumed nor required, and I presuppose only the reader's familiarity with the following simple notions: vector space, group, discrete group, ring, polynomial ring, module, associative algebra, commutative algebra, Lie algebra, derivation, graded space, $\mathbf{Z}_2$-graded space, commutative superalgebra. (The last three can be quickly learned from the introduction to [Ka 2].) Two suggestions: 1) To smooth the path through the text, the reader unfamiliar with the discrete calculus of variations may wish on the first reading to disregard all the indices pertaining to the discrete group G; 2) If the reader will find it difficult to digest the proof of the classification Theorem 5.36 in Section 5, he may wish to look up first the original source [W1] dealing with the even case.

Overall, the text should be suitable for: a beginning graduate student since *everything in the text is proven*; the specialist who has come to realize the need for a general point of view on integrable systems beyond a few celebrated but unrepresentatively special equations; and for the physicist desirous to equip himself with the basics of the superHamiltonian formalism essential for handling dynamical problems in supersymmetric situations.

Since this book concerns itself only with basic matters, some interesting advanced topics have been left out. Among the most important omissions are: 1) Nonclassical superintegrable systems ([Ku 5, 8]), i.e., those with Lax operators of the form $L = \sum_{i=0}^{n} u_i \xi^i + \sum_{j=1}^{N} \rho_j \xi^{-1} \psi_j$, where $\xi = d/dx$, u_i's, ρ_j's and ψ_j's are matrices, with ρ_j and ψ_j having equal $\mathbf{Z}_2$ degrees for each j. A new feature here is that the form of L precludes a purely elementary treatment since the corresponding Lax equations are not *a priori* self-consistent; 2) A super analog of the AKNS scheme ([G-O]) based on the Lie superalgebra $b(0, 1)$ ([Ka 2]), i.e., the unique nontrivial Lie superalgebra whose even part is $sl(2)$. A new feature here is that the highest term of the Lax operator $L = u_1 \xi + u_0$, is semisimple but *not invertible*, and therefore does not fall under the jurisdiction of the theory of classical superintegrable systems. This leads to numerous complications, not least among them being the absence of a proof that the corresponding Lax equations *exist* as local objects; 3) Supersymmetric Lax equations based on scalar Lax operators of the form $L = \sum_{i=0 \text{ or } -\infty}^{n} u_i \theta^i$, where either $n \in 1 + 2\mathbf{Z}_+$ ([Ma-R]) or $n \in 2\mathbf{N}$ ([Ku 12]), and θ is an *odd* derivation (roughly speaking $\theta = \sqrt{d/dx}$). There is a variety of new and extraordinary features here, e.g., that the conservation laws, the τ-function, and the associated Poisson brackets can all be *odd*.

I should like to acknowledge those who have helped me to see this book out. I am grateful to George Wilson for help with the proof of Theorem 6.34, to Victor Kac for suggesting the idea of this book, to Ian Stewart and Michiel Hazewinkel for inducing me to finish it, and to Arthur Greenspoon and Darryl Holm for suggesting improvements in the manuscript. I owe a debt of gratitude to my colleagues at the Space Institute for providing a stimulating environment, and especially to John Carruthers, Horace Crater, K.C. Reddy, and Susan Wu for their encouragement which is very much appreciated. I thank Sue Crawford and Linda Williams for efficiently typing the manuscript. Finally, the support of the National Science Foundation during the course of this work is gratefully acknowledged.

Logical Dependence

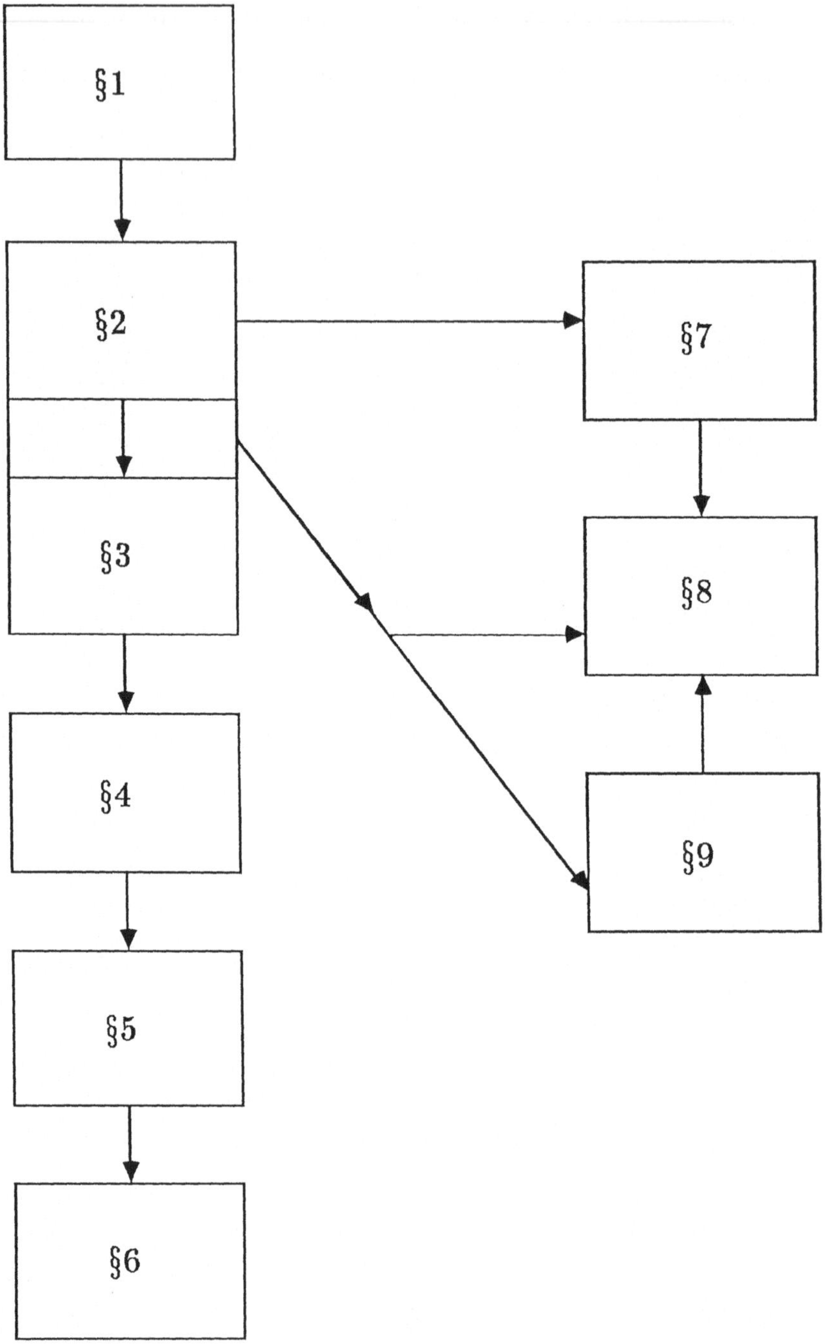

CHAPTER I

CLASSICAL SUPERINTEGRABLE SYSTEMS

§1. Introduction

In this Chapter we construct classical superintegrable systems associated to matrix pseudo-differential operators and establish their basic properties: commutativity of the flows, existence of an infinity of conservation laws, and a superHamiltonian structure.

We begin to explain this by first summarizing the basic facts in the theory of the classical integrable systems (see [W 1]). The objects of this theory are Lax equations

$$(1.1)\quad L_t = [P_+, L] = [-P_-, L],$$

where L is a fixed matrix differential operator

$$(1.2)\quad L = \sum_{i=0}^{n} u_i \xi^i, \quad n \in \mathbf{N}, \xi \equiv \partial/\partial x,$$

satisfying the following two conditions:

(1.3i) u_n is a constant, invertible, diagonalizable matrix with coefficients in a field k of characteristic zero (say, $k = \mathbf{C}$);

(1.3ii) $u_{n-1} \in Im\, ad\, u_n$.

If u_n is already diagonal, which we shall assume is the case in the main body of the paper (I will explain in a moment why this assumption is harmless),

$$(1.4i)\quad u_n = diag\,(c_1, \ldots, c_\ell), \quad k \ni c_\alpha \neq 0, \quad 1 \leq \alpha \leq \ell,$$

then the condition (1.3 ii) means that

$$(1.4ii)\quad u_{n-1,\alpha\beta} = 0 \text{ whenever } c_\alpha = c_\beta.$$

In particular, $u_{n-1,\alpha\alpha} = 0$. The operator P in (1.1) runs over the elements of the centralizer $Z(L)$ of L (that is, over all elements commuting with L) in the ring $Mat_\ell(\mathcal{O}_C)$ of matrix pseudo-differential operators, with

$$(1.5)\quad \mathcal{O}_C = \left\{ \sum_{j<\infty} f_j \xi^j | f_j \in C \right\},$$

where $C = k[u_{i,\alpha\beta}^{(m)}]$ is a (commutative) differential algebra, with free independent generators $\{u_{i,\alpha\beta}^{(m)} | m \in \mathbf{Z}_+; 0 \leq i < n, 1 \leq \alpha, \beta \leq \ell; u_{n-1,\alpha\beta}$ runs over a basis of root vectors in $Im\, ad\, u_n \subset Mat_\ell(k)\}$ (in the case that u_n is diagonal, this last condition becomes "$u_{n-1,\alpha\beta}$ is absent whenever $c_\alpha = c_\beta$ in (1.4 i)"), and with a derivation $\partial : C \to C$ over k acting on the generators of C by the rule $\partial(u_{i,\alpha\beta}^{(m)}) = u_{i,\alpha\beta}^{(m+1)}$; the derivation ∂ is naturally extended from C into $Mat_\ell(C)$ making the latter an associative differential algebra; the associative structure in $\mathcal{O}_C$ (and, thus, in $Mat_\ell(\mathcal{O}_C)$) is made possible by the relations

$$(1.6)\qquad \xi^m f = \sum_{\mathrm{k} \geq 0} \binom{m}{\mathrm{k}} f^{(\mathrm{k})} \xi^{m-\mathrm{k}}, \quad \forall f \in C,\ f^{(\mathrm{k})} := \partial^{\mathrm{k}}(f), m \in \mathbf{Z}\ ;$$

for an element $R = \sum r_j \xi^j \in Mat_\ell(\mathcal{O}_C)$, R_+ and R_- stand for the differential and "integral" part of R respectively: $R_+ = \sum_{j \leq 0} r_j \xi^j$, $R_- = \sum_{j<0} r_j \xi^j$; also, the residue of R is defined as $Res\, R := r_{-1}$.

Since P commutes with L, the two sides in (1.1) are indeed equal. The first equality in (1.1) shows that L_t is a differential operator, and the second one shows that L_t has an order $\leq n-1$ with its ξ^{n-1}-coefficient satisfying (1.3ii). Thus, the equation (1.1) can be, and is, considered as defining an evolution derivation X_P of C (see [Ku 2; Man] for the origin of this definition), that is, a derivation of C over k which commutes with ∂. In particular, any evolution derivation is uniquely defined by its actions on the generators $u_{i,\alpha\beta} = u^{(0)}_{i,\alpha\beta}$; if one thinks of $u_{i,\alpha\beta}$'s as functions of x and t, with ∂ being $\partial/\partial x$, then the equation (1.1) can be thought of as a "flow" in an appropriate function space, - both linguistic constructions are in use. Finally, to dispose of the general condition (1.3i) in favor of (1.4i), notice that if a matrix $S \in Mat_\ell(k)$ conjugates u_n from (1.3i) into its diagonal form (1.4i) then the same S, by conjugating both L and $Z(L)$, conjugates the Lax equation (1.1) into itself but with u_n being now diagonal. Thus, no loss of generality occurs in passing to (1.4) from (1.3).

With the help of the following grading w on C, $\mathcal{O}_C$, and $Mat_\ell(\mathcal{O}_C)$,

$$(1.7) \quad \mathrm{w}(u^{(m)}_{i,\alpha\beta}) = m+n-i; \mathrm{w}(k) = 0; \quad \mathrm{w}(\partial) = \mathrm{w}(\xi) = 1;$$
$$\mathrm{w}(r) = \mathrm{w}(r_{\alpha\beta}), r \in Mat_\ell(\mathcal{O}_C),$$

we can recall the main results [W 1] about Lax equations (1.1):

(1.8i) Each w-homogeneous element $P = \sum_{-\infty}^{m} p_j \xi^j \in Z(L)$ is uniquely defined by its highest term p_m which must be constant (i.e., $p_m \in Mat_\ell(k)$) and to belong to the center of $\{Z(u_n)$ in $Mat_\ell(k)\}$ (in the case u_n is diagonal this means that p_m is diagonal with $p_{m,\alpha\alpha} = p_{m,\beta\beta}$ when $c_\alpha = c_\beta$); $Z(L)$ consists of sums of its w-homogeneous generators; $Z(L)$ is abelian.

(1.8ii) For $P, Q \in Z(L)$, we have $[X_P, X_Q] = 0\,; X_P(Q) = [P_+, Q] = = [-P_-, Q]$.

(1.8iii) All the equations (1.1) have a common infinite set of conservation laws (= $c.l.$'s) $\{tr\, Res\, Q | Q \in Z(L)\}$, where a $c.l.$ for (1.1) is an element $H \in C$ such that $X_P(H) \in \{Im\,\partial$ in C $\}$.

In addition ([K - W]):

(1.9) All systems (1.1) are Hamiltonian.

(A precise explanation of the term 'Hamiltonian' is given in §3.)

Most (but not all : see e.g., [Ku 10]) integrable systems of Lax type, including those associated to Kac-Moody Lie algebras (see [D-S 1, 2; W 3]]), have either the classical form (1.1)–(1.3), or are specializations thereof, or have as their Lax operator L not a differential but a pseudo-differential operator:

$$(1.10)\quad L = \sum_{-\infty}^{n} u_i \xi^i .$$

In particular, when L is a scalar operator (i.e., $\ell = 1$) and $n = 1$, (1.10) can be written as

$$(1.11)\quad L = \xi + \sum_{i=0}^{\infty} A_i \xi^{-i-1} .$$

which is the Lax operator of the important Kadomtsev-Petviashvili (KP) hierarchy.

For operators (1.10) with the same normalization conditions (1.3), all the basic properties (1.8) of the equations (1.1) remain true [W 2]. The property (1.9) also remains true although the derivation of it, given in [G - Do 2; K - W], must be generalized, and this will be done below in §6.

The word 'super' in the title refers to the presence of anticommuting variables conveniently called 'fermions'; commutative variables are then called 'bosons'. (This language has come from physics where also the concept of supersymmetry had originated.) Our goal is to generalize classical integrable systems by allowing matrix elements in the matrices u_i's of the operator L in (1.2) [or (1.10)] to belong to a commutative superalgebra, and by requiring L

to be an even element in a fixed matrix superstructure. (See, e.g., [Ka 2; Le] for a quick introduction to linear superalgebra.) We call such systems classical superintegrable. To develop a basic theory of classical superintegrable systems, i.e., to find analogs of the properties (1.8*i* - *iii*) and (1.9), the familiar machinery of commutative differential algebras is useless. One needs instead to develop: an algebraic calculus of variations over supercommutative differential algebras, which is done in §2; superHamiltonian formalism in general and the formalism of "dual spaces" of functional Lie superalgebras in particular, which is done in §3; manipulation formulae to handle the supertrace of the Residue of the differential of a power of a matrix pseudo-differential operator, which is done in §4. Then we proceed, in §5, to describe the structure of classical superintegrabls systems and their conservation laws, generalizing properties (1.8*i* - *iii*) above; the semisimplicity conditions (1.3) and (1.4) are suitably generalized as well. Finally, in §6, we derive formulae for the variational derivatives of conservation laws, and use these formulae to derive the superHamiltonian structures of classical superintegrable systems; for the expert reader, let me mention new features there which extend and simplify even the commutative results: the semisimple matrix u_n in L is allowed to be irregular, i.e., to have multiple eigenvalues; the variables u_i's with $i \geq 0$ and with $i < 0$ split in two separate superHamiltonian structures; the corresponding superHamiltonian matrices $B_+ = B_{\geq 0}$ and $B_- = B_{<0}$ are determined *without* any computations with binomial coefficients: instead, we identify them with appropriate differential Lie superalgebras; in the checking of the superHamiltonian character of the matrix B_+, the tedious verification that the constant part of B_+ is an appropriate two-cocycle (as was done in [G - Do 3]) is avoided by a specialization trick.

Except for a few basic notions about commutative superalgebras and free modules over them, this Chapter is self-contained. No knowledge of the theory of commutative integrable systems is assumed.

§2. Variational Calculus with Anticommuting Variables

In this section we introduce the language of the differential-difference calculus and study basic grammatical constructions of this language. The main objects are: evolution fields; differential forms; variational derivatives; Fréchet derivatives; adjoint operators. The main results are: formula for the first variation; transformation properties of functional derivates; full description of both the Kernel and the Image of the variational operator.

Let k be a commutative ring with unity and K a commutative superalgebra over k, also with unity (see [Ka 2]). Let G be a discrete group acting by even automorphisms on K/k and let $\partial_1, ..., \partial_m : K/k \to K/k$ be even left derivations of K over k which commute between themselves and with the action of G. Let $I_{\bar{0}}$ and $I_{\bar{1}}$ be two disjoint countable sets, $I = I_{\bar{0}} \cup I_{\bar{1}}, N = |I|$. Let $C = C_q = K[q_i^{(g|\nu)}], i \in I, g \in G, \nu \in \mathbf{Z}_+^m$, be a commutative superalgebra with the $\mathbf{Z}_2$-grading ([Ka 2]) p on C defined by: $p(q_i^{(g|\nu)}) = p(i)$, where $p(i) = \bar{0} \in Z_2$ for $i \in I_{\bar{0}}$, and $p(i) = \bar{1} \in \mathbf{Z}_2$ for $i \in I_{\bar{1}}$. The actions of G (by automorphisms) and ∂'s (by derivations) are naturally extended on $C : \widehat{h}(q_i^{(g|\nu)}) = q_i^{(hg|\nu)}, h \in G$; and $\partial^\sigma(q_i^{(g|\nu)}) = q_i^{(g|\sigma+\nu)}$, where $(\pm\partial)^\sigma := (\pm\partial_1)^{\sigma_1}...(\pm\partial_m)^{\sigma_m}$ for $\sigma = (\sigma_1, \ldots, \sigma_m) \in \mathbf{Z}_+^m$. These actions still commute. (An informal model: $K = \oplus_G(C^\infty(M) \otimes \Lambda(r))$, where M is a manifold with coordinates $(x_1, ..., x_m)$, and $\Lambda(r)$ is a Grassmann algebra with r generators. G acts on K by moving summands according to the left action of G on itself; $\partial_s = \partial/\partial x_s$. Then $q_i^{(g|\nu)}$'s are 'coordinates' on the infinite tower of the jet bundles whose fiber coordinates are q_i's. (See [Ku 1; Man]).) Let $Der(C)$ be the Lie superalgebra of all left derivations of C over K. A Lie superalgebra $D^{ev} = D^{ev}(C)$ consists of those derivations which commute with the actions of G and ∂'s: they are called *evolution* derivations (or vector fields) ([Ku 1; Man]). If $\partial/\partial q_i^{(g|\nu)}$ denotes the natural left derivation of C/K, of the $\mathbf{Z}_2$-grading $p(i)$, then every evolution field X can be uniquely written as $X = \sum \widehat{g}\partial^\nu(X_i)\cdot \partial/\partial q_i^{(g|\nu)}$, where $X_i = X(q_i)$ and $q_i := q_i^{(e|0)}$(e is the unit element in G).

Let $\Omega^1 = \Omega^1(C) = \{\sum dq_i^{(g|\nu)}\varphi_i^{g|\nu} | \varphi_i^{g|\nu} \in C$, finite sums$\}$ be the right $\mathbf{Z}_2$-graded C-module of 1 - forms on C and $\Omega_0^1 = \Omega_0^1(C) = \{\sum dq_i\varphi_i | \varphi_i \in C$, finite sums$\}$ be the right $\mathbf{Z}_2$-graded C-module of reduced forms. The $\mathbf{Z}_2$-grading on Ω^1 is given by $p(dq_i^{(g|\nu)}) = p(i) + \bar{1}$. Let $d : C \to \Omega^1(C)$ be the universal left *odd* derivation of C over K defined by the rule $d(q_i^{(g|\nu)}) = dq_i^{(g|\nu)}$: $d(H) = \sum dq_i^{(g|\nu)}\partial H/\partial q_i^{(g|\nu)}$. The actions of $D^{ev}(C), G$ and ∂'s are uniquely extended on $\Omega^1(C)$ such that they mutually commute between themselves and with d.

Denote $Im\mathcal{D} = \sum_{g\in G} Im(\widehat{g} - \widehat{e}) + \sum_s Im\, \partial_s$. Elements of $Im\mathcal{D}$ are called trivial. We write $a \sim b$ if $(a - b) \in Im\mathcal{D}$, and say that a is equivalent to b. We define the projection $\widehat{\delta} : \Omega^1 \to \Omega_0^1$ by the rule

$$(2.1)\quad \widehat{\delta}(dq_i^{(g|\nu)} f) = dq_i(-\partial)^\nu \widehat{g}^{-1}(f).$$

Obviously ('integrating by parts'),

$$(2.2)\quad \widehat{\delta}(\omega) \sim \omega, \qquad \forall \omega \in \Omega^1.$$

The Euler-Lagrange operator $\delta : C \to \Omega_0^1(C)$ is defined as

$$(2.3)\quad \delta = \widehat{\delta}d : \delta(H) = \sum dq_i \frac{\delta H}{\delta q_i},$$

where, by (2.1),

$$(2.4)\quad \frac{\delta H}{\delta q_i} = \sum (-\partial)^\nu \widehat{g}^{-1}\left(\frac{\partial H}{\partial q_i^{(g|\nu)}}\right).$$

From now on we assume that the algebra C, as a K-module, is infinite-dimensional. This happens when one of the following conditions is satisfied:

(2.5i) There is some derivation present acting on K;

(2.5ii) There are no derivations present, but there exists at least one even variable among the q's;

(2.5iii) There are no derivations and no even q's, but the number $N = |I|$ (of odd q's) is infinite;

(2.5$iiii$) There are no derivations and no even q's, the number $N = |I|$ is finite, but the group G is infinite.

It follows that $C_{\bar{0}}$, as a $K_{\bar{0}}$ -module, is also infinite-dimensional.

2.6. Proposition. $\hat{\delta}(Im\mathcal{D}) = 0$.

Proof. (a) $\hat{\delta}\,\hat{h}(dq_i^{(g|\nu)} f) = \hat{\delta}\left[dq_i^{(hg|\nu)}\,\hat{h}\,(f)\right] =$

$= dq_i(-\partial)^\nu((\widehat{hg})^{-1}\hat{h}(f) = dq_i(-\partial)^\nu\,\hat{g}^{-1}(f) = \hat{\delta}(dq_i^{(g|\nu)} f) =$

$= \hat{\delta}\,\hat{e}\,(dq_i^{(g|\nu)} f)$. Thus, $\hat{\delta}(\hat{h} - \hat{e}) = 0,\ \forall\, h \in G$;

(b) $\hat{\delta}\partial_s(dq_i^{(g|\nu)} f) = \hat{\delta}\,[dq_i^{(g|\nu+1_s)} f + dq_i^{(g|\nu)}\partial_s(f)] =$

$= dq_i\,\hat{g}^{-1}\,[(-\partial)^{\nu+1_s}(f) + (-\partial)^\nu\partial_s(f)] = dq_i\,\hat{g}^{-1}(-\partial)^\nu\,[-\partial_s(f) + \partial_s(f)] = 0.$

Thus, $\hat{\delta}\partial_s = 0$. ∎

2.7. Corollary. $\dfrac{\delta}{\delta q_i}(Im\mathcal{D}) = 0.$

Proof. $\sum dq_i \dfrac{\delta H}{\delta q_i} = \delta(H) = \hat{\delta}d(H)$, and if $H \in Im\mathcal{D}$ then $d(H) \in Im\mathcal{D}$, and hence $\hat{\delta}d(H) = 0$ by Proposition 2.6. ∎

The following result serves the role played by the DuBois-Reymond Lemma in the traditional calculus of variations.

2.8. Lemma. If f is an element in C and $f C_{\bar{0}} \sim 0$ then $f = 0$.

Proof. (i) Let us consider first the case (2.5i) when $m > 0$, that is, when there is some derivation present. Let $M \in \mathbf{Z}_+$ be such that $\dfrac{\partial f}{\partial q_i^{(g|\nu)}} = 0$ for $\forall \nu$ with $\nu_m \geq M$ and for some $i \in I$. We have, by Corollary 2.7,

(a) If $p(i) = \bar{0}$ then $0 = \dfrac{\partial}{\partial q_i^{(e|2M_m)}} \dfrac{\delta}{\delta q_i} \left[(q_i^{(e|M_m)})^2 f \right] = (-1)^M 2f$;

(b) If $p(i) = \bar{1}$ then $0 = \dfrac{\partial}{\partial q_i^{(e|(2M+1)_m)}} \dfrac{\delta}{\delta q_i} \left[q_i^{(e|M_m)} q_i^{(e|(M+1)_m)} f \right] =$

$= (-1)^M 2f.$

Thus, $f = 0$.

Now assume that there are no derivations present to begin with. Let the degree of f be $\leq M$. We consider separately the cases ($2.5ii - iiii$).

(ii) Suppose that there exists some even variable $q = q_i$ (case (2.5ii)). Choose $\varphi = q^{M+\mathrm{k}+1}$ with some $\mathrm{k} \in \mathbf{Z}_+$. Then

$$0 = \frac{\delta}{\delta q}(\varphi f) = \sum_{g \neq e} \hat{g}^{-1} \left[\varphi \frac{\partial f}{\partial q^{(g)}} \right] + \frac{\partial}{\partial q}(\varphi f).$$

Applying to this equality the operator $\left(\dfrac{\partial}{\partial q} \right)^M$ we get

$$0 = \left(\frac{\partial}{\partial q} \right)^{M+1} (q^{M+\mathrm{k}+1} f),$$

and since k is arbitrary we conclude that $f = 0$;

(iii) Suppose that there exists an unlimited number of odd variables (case (2.5iii)). Choose $i,\ j \in I, i \neq j$, such that $\frac{\partial f}{\partial q_i^{(g)}} = \frac{\partial f}{\partial q_j^{(g)}} = 0,\ \forall\, g \in G$.

Since $f q_i q_j \sim 0$,

$$0 = \frac{\delta}{\delta q_j}\frac{\delta}{\delta q_i}(q_i q_j f) = f,$$

so that $f = 0$;

$(iiii)$ Finally, suppose that there exists only a finite number of q's, all of them odd, but the group G is infinite (case (2.5$iiii$)). Assume for the moment that $f \neq 0$. We start with two remarks. First, if $fC_{\bar{0}} \sim 0$ then every homogeneous component of f (with respect to the usual degree) satisfies the same property so we may assume that f is degree-homogeneous. Second, if f is such that $fC_{\bar{0}} \sim 0$, then, for every $\psi \in C_{\bar{0}}$, $f\psi$ satisfies the same property. We choose ψ such that $f\psi$ becomes a (non-zero) monomial. To achieve that, pick some monomial $\tilde{f}$ in f. (If $f \in K$, set $\tilde{f} = f$.) Let $\tilde{\psi}$ be a product (in some order) of all those $q_i^{(g)}$'s that are present in f but absent in $\tilde{f}$. (If there are no such $q_i^{(g)}$'s, then $f = \tilde{f}$ and we take $\psi = 1$.) If $\tilde{\psi}$ is even, we set $\psi = \tilde{\psi}$. If $\tilde{\psi}$ is odd, we let $\psi = \tilde{\psi}\, q_i^{(g)}$ with some $q_i^{(g)}$ not found among monomials in f. The resulting $\bar{f} = f\psi = \tilde{f}\psi$ is now a monomial, and $\bar{f}C_{\bar{0}} \sim 0$. We consider two possibilities: (a) $\bar{f} \in K$ and (b) $\bar{f} \notin K$.

(a) If $\bar{f} \in K$, then for some $i \in I$ and $\forall\, g \in G, g \neq e$, we have

$$0 = \frac{\delta}{\delta q_i}[q_i q_i^{(g)} \bar{f}] = q_i^{(g)}\bar{f} - q_i^{(g^{-1})}\hat{g}^{-1}(\bar{f}),$$

and since $\overline{f} \neq 0$, we conclude that $g = g^{-1}, \forall g \in G$, and also $\widehat{g}(\overline{f}) = \overline{f}$, $\forall g \in G$. Hence, for any $g, h \in G$, $gh = (gh)^{-1} = h^{-1}g^{-1} = hg$, which implies that G is abelian. Let $x \in G$, $x \neq e$. Since $x^2 = e, x^{-1} = x$, and G is infinite, there exists $y \in G, y \neq x, y \neq e$. Then $G' = \{e, x, y, xy\}$ is a four-element subgroup in G. Again, since G is infinite, we can find $z \in G$, such that $z \notin G'$. It follows that $xz \notin G', yz \notin G'$. Now set $\Theta = q\, q^{(x)}\, q^{(y)}\, q^{(z)}$ with some $q = q_i, i \in I$. Since G acts identically on $\overline{f}$, we have

$$0 = \frac{\delta}{\delta q}(\Theta \overline{f}) = \frac{\delta}{\delta q}(\Theta) \cdot \overline{f}.$$

Now,

$$\frac{\delta}{\delta q}(\Theta) = q^{(x)}\, q^{(y)}\, q^{(z)} - q^{(x)}\, q^{(xy)}\, q^{(xz)} + q^{(y)}\, q^{(yx)}\, q^{(yz)} - q^{(z)}\, q^{(zx)}\, q^{(zy)}.$$

Since neither of the elements xy, xz, yz belongs to the set $\{x, y, z\}$, among the four monomials above only one has as its exponents the set $\{x, y, z\}$. Thus, applying to the equality $0 = \frac{\delta}{\delta q}(\Theta)\overline{f}$ the operator $\frac{\partial}{\partial q^{(z)}} \frac{\partial}{\partial q^{(y)}} \frac{\partial}{\partial q^{(x)}}$, we obtain $\overline{f} = 0$, a contradiction.

(b) It remains to consider the case $\overline{f} \notin K$. Pick some $i \in I$ such that $q_i^{(g)}$ is present in $\overline{f}$ for some $g \in G$. Set $q = q_i$.

Let $G_1 = \{g \in G | \frac{\partial \overline{f}}{\partial q^{(g)}} \neq 0\}$. Notice that if $\overline{f}C_0 \sim 0$ then $\widehat{g}(\overline{f})C_0 \sim 0$, $\forall g \in G$. We, thus, can assume that $e \notin G_1$. Set $G_2 = \{g = g_1^{-1}g_2 | g_1, g_2 \in G_1\}$. Choose $h \in G$ such that $h \notin \{G_2 \cup G_2^{-1} \cup G_1 \cup G_1^{-1}\}$. Now consider $\theta = \overline{f}q^{(h)}q$.

Let us write $\theta = q^{(g_1)} ... q^{(g_n)} q^{(h)} q\, c$, where $c \in C$ is such that $\frac{\partial c}{\partial q^{(g)}} = 0$, $\forall g \in G$, We have

$$(2.9)\quad 0=\frac{\delta}{\delta q}(\theta)=\sum_{s=1}^{n}(-1)^{s+1}\widehat{g_s^{-1}}\frac{\partial}{\partial q^{(g_s)}}(\theta)+$$
$$+(-1)^n q^{(h^{-1}g_1)}\dots q^{(h^{-1}g_n)}q^{(h^{-1})}\widehat{h^{-1}}(c)+(-1)^{n+1}q^{(g_1)}\dots q^{(g_n)}q^{(h)}c.$$

I claim that the set of exponents (of q's) of the last monomial in the right-hand-side of (2.9) is not matched by the set of exponents of any other monomial. To see that this is so, let us look first at the exponents of the each of the first n terms. In the s-th term, they are $\{g_s^{-1}g_j, j\neq s; g_s^{-1}h;\ g_s^{-1}\}$. The way we have chosen h, there is no exponent h in the latter set. Since h is present in the last monomial in (2.9), the only way for the right-hand-side of (2.9) to vanish is for the last two terms in (2.9) to cancel each other out. In particular, these two terms must have the same set of exponents. Since the first of these terms has h^{-1} and $h^{-1}\notin G_1=\{g_1,\dots,g_n\}$, we must have $h=h^{-1}$. But then $h^{-1}g_i(=hg_i)$ in the first of these two terms has to match some g_j in the second of these terms. This forces $h=g_jg_i^{-1}$ which contradicts the way we chose h. The only possibility left for (2.9) to vanish must reside at the choice of c : c must be zero. Hence, $\overline{f}=0$, a contradiction. ∎

2.10. <u>Definition.</u> The natural pairing $< Der(C), \Omega^1(C) >$ is defined by the formula

$$<\sum Y_i^{g|\nu}\frac{\partial}{\partial q_i^{(g|\nu)}},\sum dq_j^{(h|\sigma)}\varphi_j^{h|\sigma}>=\sum Y_i^{g|\nu}\varphi_i^{g|\nu}.$$

2.11. <u>Remark.</u> The order in which the pairing is taken is important: $Der\,(C)$ is a left C-module, while $\Omega^1\,(C)$ is a right C-module. (In the absence of odd variables the order makes, of course, no difference.)

2.12. <u>Lemma.</u> If $\omega\in\Omega_0^1\,(C)$ and $< Der\,(C),\omega >\sim 0$ then $\omega=0$.

<u>Proof.</u> If $\omega\neq 0$ then there exists $Z\in Der\,(C)$ such that $< Z,\omega >=:$

$f \neq 0$. Then, for any $\varphi \in C$, $< \varphi Z, \omega > = \varphi f \sim 0$. By Lemma 2.8, $f = 0$. A contradiction. ∎

The classical calculus of variations operates with 'variations'. We are now ready to establish the relations between the two approaches.

2.13. Theorem. (a) If $\omega \in \Omega_0^1$ and $\omega \sim 0$ then $\omega = 0$; (b) If $\omega \in \Omega^1$ then $\omega \sim 0$ if and only if $< D^{ev}, \omega > \sim 0$; (c) The projection $\widehat{\delta} : \Omega^1 \to \Omega_0^1$ can be uniquely defined by the formula $< X, \widehat{\delta}(\omega) > \sim < X, \omega >$, $\forall X \in D^{ev}$; (d) $Ker\,\widehat{\delta} = Im D$ in Ω^1.

Proof. The 'if' part of (b): For $X \in D^{ev}$, we have

$$< X, (\widehat{h} - \widehat{e})(dq_i^{(g|\nu)} f) > = < X, dq_i^{(hg|\nu)} \widehat{h}(f) - dq_i^{(g|\nu)} f > =$$

$$= (\widehat{hg})\partial^\nu(X_i) \cdot \widehat{h}(f) - \widehat{g}\partial^\nu(X_i) f = (\widehat{h} - \widehat{e})[\widehat{g}\partial^\nu(X_i) \cdot f], \text{ and also}$$

$$< X, \partial_s(dq_i^{(g|\nu)} f) > = < X, dq_i^{(g|\nu + 1_s)} f + dq_i^{(g|\nu)} \partial_s(f) > = \widehat{g}\partial^\nu \partial_s(X_i) f +$$

$+\widehat{g}\partial^\nu(X_i)\partial_s(f) = \partial_s[\widehat{g}\partial^\nu(X_i) \cdot f]$, which proves the 'if' part of (b): $\omega \sim 0 \Rightarrow$ $< D^{ev}, \omega > \sim 0$; (a) follows from the 'if' part of (b) and Lemma 2.12; (c): Uniqueness follows from Lemma 2.12, and existense follows from (2.2) and the 'if' part of (b); (d): By Proposition 2.6, $Im D \subset Ker\,\widehat{\delta}$. On the other hand, if $\widehat{\delta}(\omega) = 0$ then $\omega \sim \widehat{\delta}(\omega) = 0$, so that $\omega \in Im D$. In other words, $Ker\,\widehat{\delta} \subset Im D$. Thus, $Ker\,\widehat{\delta} = Im D$ in Ω^1; The 'only if' part of (b): Let $< D^{ev}, \omega > \sim 0$. Since $\omega = \widehat{\delta}(\omega) + (1 - \widehat{\delta})(\omega)$ and $(1 - \widehat{\delta})(\omega) \sim 0$ by (2.2), $< D^{ev}, (1 - \widehat{\delta})(\omega) > \sim 0$ by the 'if' part of (b). Thus, $< D^{ev}, \widehat{\delta}(\omega) > \sim 0$. Hence, $\widehat{\delta}(\omega) = 0$ by Lemma 2.12. Therefore, $\omega \sim 0$ by (d). ∎

2.14. Remark. By Corollary 2.7, $Im D \subset Ker\,\delta$ in C. In fact, $Ker\,\delta = Im D + K$ in C. (See Remark 2.75 for the Proof.)

We shall often use Theorem 2.13 in the form (2.17) below, which can be called the (weak) formula for the first variation (cf. [Ku 2]). For $X \in D^{ev}$, $H \in C$, we denote

$$(2.15) \quad \overline{X} = (X_i)\,,\; X_i := X(q_i),$$

$$(2.16) \quad \frac{\delta H}{\delta \overline{q}} = \left(\frac{\delta H}{\delta q_i}\right),$$

the corresponding column vectors. Then, $\forall\, X \in D^{ev}$,

$$(2.17) \quad X(H) = < X, d(H) > \sim < X, \delta(H) > = \sum X_i \frac{\delta H}{\delta q_i} = \overline{X}^t \frac{\delta H}{\delta \overline{q}},$$

and this relation uniquely defines the vector $\dfrac{\delta H}{\delta q}$.

We now turn to the natural properties of the calculus of variations. We begin with a few useful technicalities.

2.18. Definition. Let $R = (R_r | R_r \in C)$ be a column vector. Its commutative Fréchet derivative $D(R)\,(= D_q(R))$ is a matrix operator with the matrix elements

$$(2.19) \quad [D(R)]_{ri} = D_i(R_r) = \sum \frac{\partial R_r}{\partial q_i^{(g|\nu)}} \hat{g}\,\partial^\nu$$

2.20. Definition. Let $f \in C$ be a $\mathbf{Z}_2$–homogeneous element. Its even and odd Fréchet derivatives are the following row vectors of operators:

$$(2.21) \quad [D^{\overline{0}}(f)]_i = D_i^{\overline{0}}(f) = \sum (-1)^{p(i)[p(f)+\overline{1}]} \frac{\partial f}{\partial q_i^{(g|\nu)}} \hat{g}\,\partial^\nu,$$

$$(2.22) \quad [D^{\overline{1}}(f)]_i = D_i^{\overline{1}}(f) = \sum (-1)^{p(f)[p(i)+\overline{1}]} \frac{\partial f}{\partial q_i^{(g|\nu)}} \hat{g}\,\partial^\nu,$$

where we use the usual notation

$$(2.23) \quad (-1)^{\overline{0}} = 1,\; (-1)^{\overline{1}} = -1.$$

We shall often drop the bar off the elements $\bar{0}, \bar{1} \in \mathbf{Z}_2$, to simplify the notation. The Fréchet derivatives have the following property.

2.24. Lemma. Let $X \in D^{ev}$ be $\mathbf{Z}_2$-homogeneous, and let $f \in C$ be $\mathbf{Z}_2$-homogeneous as well. Then $X(f) = D^{p(X)}(f)(\overline{X})$.

Proof. Recall that the usual grading of operators mapping one graded space into another is defined as the difference between the gradings of image and preimage. In particular,

$$(2.25) \quad p(X) = p(X_i) - p(q_i) = p(X_i) + p(i).$$

Now,

$$\begin{aligned} X(f) &= < X, d(f) > = \sum \widehat{g}\partial^{\nu}(X_i) \cdot \frac{\partial f}{\partial q_i^{(g|\nu)}} = \\ &= \sum (-1)^{[p(X)+p(i)][p(f)+p(i)]} \frac{\partial f}{\partial q_i^{(g|\nu)}} \widehat{g}\partial^{\nu}(X_i) = \\ &= \begin{cases} D^0(f)(\overline{X}), \text{ for } p(X) = 0\,, \\ D^1(f)(\overline{X}), \text{ for } p(X) = 1\,, \end{cases} \end{aligned}$$

since $p(i)\,[p(f) + p(i)] = p(i)[p(f) + 1]$ and $[1 + p(i)][p(f) + p(i)] =$ $= [p(i) + 1]p(f)$, and these are exactly the exponents in (2.21), (2.22). ∎

The use of the commutative Fréchet derivative D will be made later on.

If $R = (R_r | R_r \in C)$ is a vector, then the even and odd Fréchet derivatives of R are defined component-wise.

$$(2.26) \quad [D^{\gamma}(R)]_{ri} = [D^{\gamma}(R_r)]_i, \quad \forall \gamma \in \mathbf{Z}_2.$$

Also, every $X \in D^{ev}$ acts on the vector R component-wise:

$$(2.27) \quad [X(R)]_r = X(R_r).$$

2.28. Definition. Let $R = (R_i | R_i \in C, i \in I)$ be a $\mathbf{Z}_2$-homogeneous vector. Its even and odd Fréchet derivatives are the following matrix operators:

$$(2.29) \quad [D^0(R)]_{ij} = D^0_j(R_i) = \sum (-1)^{p(j)[p(R)+p(i)+1]} \frac{\partial R_i}{\partial q_j^{(g|\nu)}} \widehat{g}\, \partial^\nu,$$

$$(2.30) \quad [D^1(R)]_{ij} = D^1_j(R_i) = \sum (-1)^{[p(j)+1][p(R)+p(i)]} \frac{\partial R_i}{\partial q_j^{(g|\nu)}} \widehat{g}\, \partial^\nu.$$

2.31. Corollary. If $X \in D^{ev}$ and R are both $\mathbf{Z}_2$-homogeneous, then

$$(2.32) \quad X(R) = D^{p(X)}(R)(\overline{X}).$$

Proof, of course, follows from Lemma 2.24 and formula (2.26). ∎

2.33. Definition. An operator $A : C^r \to C^s$ is a k-linear map of the form

$$(2.34) \quad [A(v)]_a = \sum_b A_{ab}(v_b),$$

$$(2.35) \quad A_{ab} = \sum A_{ab}^{g|\nu} \widehat{g}\, \partial^\nu, \quad \text{finite sum}, \ A_{ab}^{g|\nu} \in C,$$

where both C^r and C^s are considered consisting of column vectors. (Although δ is not an operator in this sense, we shall continue to call it the Euler-Lagrange operator, respecting the tradition.)

2.36. Definition. For a $\mathbf{Z}_2$-homogeneous operator $A : C \to C$, its adjoint is an operator $A^\dagger : C \to C$ satisfying

$$(2.37) \quad [A^\dagger(u)]\, v \sim (-1)^{p(u)p(v)} [A(v)]\, u, \quad \forall v, u \in C.$$

2.38. Lemma. The adjoint operator exists and is unique.

Proof. Uniqueness follows from Lemma 2.8, and existense follows from the following computation:

$$(-1)^{p(u)p(v)}[A(v)]u = (-1)^{p(u)p(v)} A^{g|\nu}\widehat{g}\partial^{\nu}(v)\cdot u = A^{g|\nu} u\widehat{g}\partial^{\nu}(v) \sim$$
$$\sim [\widehat{g}^{-1}(-\partial)^{\nu}(A^{g|\nu}u)]\cdot v.$$

Thus,

$$(2.39)\quad (f\widehat{g}\partial^{\nu})^{\dagger} = \widehat{g}^{-1}(-\partial)^{\nu} f,\ \ f\in C.$$ ∎

2.40. Definition. Let $A = (A_{ab})$ be an operator. Its commutative adjoint , $A^{\dagger}$, is defined as

$$(2.41)\quad \left(A^{\dagger}\right)_{ba} = (A_{ab})^{\dagger}.$$

2.42. Definition. For an operator $A : C^N \to C^N, N = |I|$, its superadjoint, $A^{s\dagger}$, is defined as

$$(2.43)\quad \left(A^{s\dagger}\right)_{ji} = (-1)^{p(i)p(j)}(A_{ij})^{\dagger}.$$

2.44. Proposition. Let $u, v \in C^N$ be even vectors. Then

$$(2.45)\quad [A(u)]^t v \sim [A^{s\dagger}(v)]^t u.$$

Proof. Recall that a vector $\mathrm{w} = (\mathrm{w}_i) \in C^N$ is even (resp., odd) if $p(\mathrm{w}_i) = p(i)$ (resp., $p(\mathrm{w}_i) = p(i) + 1$), $\forall i \in I$. Now,

$[A(u)]^t v = \sum A_{ij}(u_j)\cdot v_i \sim (-1)^{p(u_j)p(v_i)}[(A_{ij})^{\dagger}(v_i)]\cdot u_j$ [by (2.43)] $=$ $[A^{s\dagger}(v)]^t u.$ ∎

Now we can tackle the transformation properties of the Euler-Lagrange operator.

Let $C_1 = K[u_j^{(g|\nu)}]$, $j \in J = J_{\bar{0}} \cup J_{\bar{1}}, g \in G, \nu \in \mathbf{Z}_+^m$, be another commutative superalgebra. Let $\Phi : C \to C_1$ be an even homomorphism over K, commuting with the actions of G and ∂'s: Such a homomorphism is called a differential-difference homomorphism, or simply a homomorphism for brevity. We uniquely extend Φ to the map $\Phi : \Omega^1(C) \to \Omega^1(C_1)$ by requiring it to commute with d and with the actions of G and ∂'s:

$$
\begin{aligned}
\Phi(dq_i^{(g|\nu)}) &= d\,\Phi(q_i^{(g|\nu)}) = d\,(\widehat{g}\,\partial^\nu\,\Phi(q_i)) = \widehat{g}\,\partial^\nu\, d(\Phi_i) = \\
&= \widehat{g}\partial^\nu \left(\sum du_j^{(h|\sigma)} \frac{\partial \Phi_i}{\partial u_j^{(h|\sigma)}} \right),
\end{aligned}
\tag{2.46}
$$

where

$$\Phi_i = \Phi(q_i). \tag{2.47}$$

In particular,

$$
\begin{aligned}
\Phi(dq_i) &= \sum du_j^{(h|\sigma)} \frac{\partial \Phi_i}{\partial u_j^{(h|\sigma)}} \sim \sum du_j\, \widehat{h}^{-1}(-\partial)^\sigma \left(\frac{\partial \Phi_i}{\partial u_j^{(h|\sigma)}} \right) = \\
&= \sum du_j \frac{\delta \Phi_i}{\delta u_j} = \delta_1(\Phi_i),
\end{aligned}
\tag{2.48}
$$

where we use subscript '1' to distinguish operations in C_1 from those in C.

2.49. Lemma $\widehat{\delta}_1\,\Phi\,\widehat{\delta} = \widehat{\delta}_1\,\Phi$.

Proof. Since Φ commutes with the action of G and ∂'s,

$$\Phi(Im\mathcal{D}) \subset Im\mathcal{D}. \tag{2.50}$$

By (2.2), $Im(\widehat{\delta} - 1) \subset Im\mathcal{D}$, hence $Im\Phi(\widehat{\delta} - 1) \subset Im\mathcal{D}$. Therefore, $\widehat{\delta}_1\,\Phi(\widehat{\delta} - 1) = 0$. ∎

Denote by $\overline{\Phi}$ the following column vector:

$$(2.51) \quad (\overline{\Phi})_i = \Phi_i = \Phi(q_i).$$

2.52. Theorem. For any $H \in C$,

$$(2.53) \quad \frac{\delta \Phi(H)}{\delta \overline{u}} = D(\overline{\Phi})^{\dagger} \Phi \left(\frac{\delta H}{\delta \overline{q}} \right).$$

Proof. We have,

$$\begin{aligned}
\sum du_j \frac{\delta \Phi(H)}{\delta u_j} &= \delta_1 \Phi(H) = \widehat{\delta}_1 d\, \Phi(H) = \widehat{\delta}_1 \Phi d(H) \text{ [by Lemma 2.4]} \\
&= \widehat{\delta}_1 \Phi \widehat{\delta} d(H) = \widehat{\delta}_1 \Phi \delta(H) = \widehat{\delta}_1 \Phi \left(\sum dq_i \frac{\delta H}{\delta q_i} \right) \text{ [by (2.48)]} \\
(2.54) \qquad &= \widehat{\delta}_1 \left(\sum du_j^{(h|\sigma)} \frac{\partial \Phi_i}{\partial u_j^{(h|\sigma)}} \Phi \left(\frac{\delta H}{\delta q_i} \right) \right) \text{ [by (2.1)]} \\
&= \sum du_j \widehat{h}^{-1} (-\partial)^{\sigma} \frac{\partial \Phi_i}{\partial u_j^{(h|\sigma)}} \Phi \left(\frac{\delta H}{\delta q_i} \right).
\end{aligned}$$

Comparing the first and the last terms in (2.54) and using Theorem 2.13 (a), we obtain

$$(2.55) \qquad \frac{\delta \Phi(H)}{\delta u_j} = \sum \widehat{h}^{-1} (-\partial)^{\sigma} \frac{\partial \Phi_i}{\partial u_j^{(h|\sigma)}} \left[\Phi \left(\frac{\delta H}{\delta \overline{q}} \right) \right]_i.$$

It remains to notice that

$$\sum \widehat{h}^{-1} (-\partial)^{\sigma} \frac{\partial \Phi_i}{\partial u_j^{(h|\sigma)}} = \sum \left(\frac{\partial \Phi_i}{\partial u_j^{(h|\sigma}} \widehat{h} \partial^{\sigma} \right)^{\dagger} \text{ [by (2.19) and (2.51)]}$$

$= [D(\overline{\Phi})_{ij}]^{\dagger}$ [by (2.41)] $= [D(\overline{\Phi})^{\dagger}]_{ji}$. ∎

Our last topic in this section is to describe the image of the Euler-Lagrange operator $\delta : C \rightarrow \Omega_0^1(C)$. To do that we first construct a complex for δ, and then show that this complex is exact in the $\Omega_0^1(C)$-place.

Denote $\overline{C} = K[q_i^{(g|\overline{\nu})}], i \in I, g \in G, \overline{\nu} \in \mathbf{Z}_+^{m+1}$, and let $\partial_{m+1} : \overline{C} \rightarrow \overline{C}$ be a new even left derivation which acts trivially on K, commutes with G and $\partial_1,...,\partial_m$, and sends $q_i^{(g|\overline{\nu})}$ into $q_i^{(g|\overline{\nu}+1_{m+1})}$. Let us write $(g|\nu|s)$ instead of $(g|\overline{\nu})$ for $\overline{\nu} = \nu \oplus s, \nu \in \mathbf{Z}_+^m, s \in \mathbf{Z}_+$, where we identified $\mathbf{Z}_+^{m+1}$ with $\mathbf{Z}_+^m \oplus \mathbf{Z}_+$. The old notation $(g|\nu)$ will be used instead of $(g|\nu|0)$. This, in fact, means that we fix an injective homomorphism $C \hookrightarrow \overline{C}$ which we suppress from notation and thereafter consider C sitting inside $\overline{C}$.

Let $\overline{\tau} : \Omega^1(C) \rightarrow \overline{C}$ be an odd homomorphism of right C-modules given as

(2.56) $\overline{\tau}(dq_i^{(g|\nu)} f) = q_i^{(g|\nu|1)} f, \quad f \in C.$

Since $\overline{\tau}$ obviously commutes with the actions of G and $\partial_1, ..., \partial_m$, we see that

(2.57) $\overline{\tau}(Im D) \subset Im D$.

2.58. Lemma. $\partial_{m+1}(H) = \overline{\tau} d(H), \qquad \forall H \in C.$

Proof. We have

(2.59) $$\overline{\tau} d(H) = \overline{\tau}\left[\sum dq_i^{(g|\nu)} \frac{\partial H}{\partial q_i^{(g|\nu)}}\right] = \sum q_i^{(g|\nu|1)} \frac{\partial H}{\partial q_i^{(g|\nu)}} = \partial_{m+1}(H).$$ ∎

Let superscript '1' in the operators δ^1 and $\widehat{\delta}^1$ refer to the algebra $\overline{C}$, so that $\delta^1 = \widehat{\delta}^1 d$.

2.60. Theorem. The sequence

(2.61) $C \xrightarrow{\delta} \Omega_0^1(C) \xrightarrow{\delta^1\overline{\tau}} \Omega_0^1(\overline{C})$

is a complex.

Proof. For any $H \in C, \delta(H) = \widehat{\delta}d(H) \sim d(H)$ by (2.2). Hence, $\overline{\tau}\delta(H) \sim \overline{\tau}d(H) = \partial_{m+1}(H)$ by (2.57) and Lemma 2.58. Therefore, $\delta^1\overline{\tau}\delta(H) = \delta^1\partial_{m+1}(H) = 0$ by Corollary 2.7. ∎

We now transform the equality $0 = \delta^1\overline{\tau}\delta$ into a more convenient form.

2.62. Definition. An operator $A : C^N \to C^N, N = |I|$, is called supersymmetric (resp. superskewsymmetric) if $A^{s\dagger} = A$ (resp. $A^{s\dagger} = -A$.)

2.63. Lemma. Let us identify $\Omega_0^1(C)$ with $C^N, N = |I| : \sum dq_i R_i \leftrightarrow \overline{R} = (R_i)$. Then the equality $\delta^1\overline{\tau}(\overline{R}) = 0$, for a $\mathbf{Z}_2$-homogeneous $\overline{R}$, is equivalent to $D(\overline{R})$ being supersymmetric:

(2.64) $D(\overline{R})^{s\dagger} = D(\overline{R})$.

Proof. We have,

$$0 = \delta^1\overline{\tau}(\sum dq_i R_i) = \delta^1(\sum q_i^{(e|0|1)} R_i) = \sum dq_j \frac{\delta}{\delta q_j}[q_i^{(e|0|1)} R_i] =$$

$$\sum dq_j \widehat{g}^{-1}(-\partial)^\nu(-\partial_{m+1})^s \frac{\partial}{\partial q_j^{(g|\nu|s)}}[q_i^{(e|0|1)} R_i] =$$

$$= \sum dq_j\{\widehat{g}^{-1}(-\partial)^\nu[(-1)^{p(j)p(i)} q_i^{(e|0|1)} \frac{\partial R_i}{\partial q_j^{(g|\nu)}}] - \partial_{m+1}(R_j)\},$$

which is equivalent to

$$\sum \widehat{g}^{-1}(-\partial)^\nu[(-1)^{p(i)p(j)}(-1)^{p(i)[p(R_i)+p(j)]} \frac{\partial R_i}{\partial q_i^{(g|\nu)}} q_i^{(e|0|1)}] =$$

$$= \partial_{m+1}(R_j) = \sum q_i^{(g|\nu|1)} \frac{\partial R_j}{\partial q_i^{(g|\nu)}} =$$

$$= \sum (-1)^{p(i)[p(R_j)+p(i)]} \frac{\partial R_j}{\partial q_i^{(g|\nu|1)}} q_i^{(g|\nu|1)} =$$

$$= \sum (-1)^{p(i)[p(R_j)+p(i)]} \frac{\partial R_j}{\partial q_i^{(g|\nu)}} \widehat{g}\partial^\nu (q_i^{(e|0|1)}),$$

and since $q_i^{(g|\nu|1)}$'s are independent, we can drop them off the above equality and to transform it into an equivalent form, using (2.39) and the formula $p(R_i) = p(\overline{R}) + p(i)$:

$$\sum_{g,\nu} (-1)^{p(i)p(j)} \left(\frac{\partial R_i}{\partial q_j^{(g|\nu)}} \widehat{g}\partial^\nu \right)^\dagger = \sum_{g,\nu} \frac{\partial R_j}{\partial q_i^{(g|\nu)}} \widehat{g}\partial^\nu .$$

With the help of (2.43) this becomes

$$\left[D(\overline{R})^{s\dagger} \right]_{ji} = [D(\overline{R})]_{ji},$$

which is (2.64). ∎

2.65. Theorem. For any $H \in C$, the commutative Fréchet derivative

$D\left(\frac{\delta H}{\delta \overline{q}}\right)$ is supersymmetric:

(2.66) $$D\left(\frac{\delta H}{\delta \overline{q}}\right)^{s\dagger} = D\left(\frac{\delta H}{\delta \overline{q}}\right).$$

Proof. By Theorem 2.60, $\delta\, {}^1\overline{\tau}(\delta H) = 0$. By Lemma 2.63, the latter equation can be rewritten in the form (2.64) which yields (2.66). ∎

We now show that the sequence (2.61) is exact.

2.67. Theorem. Let $\overline{R} \in C^N, N = |I|$, be a finite vector (i.e., with only a finite number of non-zero components) for which

$$D(\overline{R})^{s\dagger} = D(\overline{R}). \tag{2.68}$$

Then there exists $H \in C$ such that $\overline{R} = \frac{\delta H}{\delta \overline{q}}$.

Proof. Define a map $A_t : C \to C[t]$ by $A_t(f) = t^{deg(f)} f$ for degree-homogeneous elements $f \in C$, and extend it by additivity on C. Set

$$H = \sum_i q_i \int_0^1 A_t\,(R_i)\,dt\,. \tag{2.69}$$

We shall show that $\frac{\delta H}{\delta \overline{q}} = \overline{R}$. By Theorem 2.13(c) in the form (2.17), it is enough to show that $X(H) \sim \overline{X}^t \overline{R}, \forall X \in D^{ev}$. Obviously, it is enough to consider only $\mathbf{Z}_2$-homogeneous X's and $\overline{R}$'s, and degree-homogeneous $\overline{R}$'s. For those, we have

$$\begin{aligned} X(H) &= \sum X_i \int_0^1 A_t(R_i)dt + \\ &+ \sum \int_0^1 X(A_t(R_i))\,dt \cdot q_i(-1)^{p(i)p(R_i)} \end{aligned} \tag{2.70}$$

Notice that if $\overline{R}$ satisfies (2.68) then so does $A_t(\overline{R})$. Let us concentrate on the second term in (2.70). Denoting $Q_i = A_t(R_i)$, and using (2.19), (2.43), and

(2.68), we have consequently

$$\sum X(Q_i)q_i = \sum \widehat{g}\partial^\nu(X_j)\cdot\frac{\partial Q_i}{\partial q_j^{(g|\nu)}}q_i = \sum(-1)^{p(X_j)[p(Q_i)+p(j)]}D_j(Q_i)(X_j)\cdot q_i,$$

$$D_j(Q_i)(X_j)\cdot q_i = D(\overline{Q})_{ij}(X_j)\cdot q_i \sim [D(\overline{Q})_{ij}]^\dagger(q_i)\cdot X_j(-1)^{p(i)p(X_j)} =$$

(2.71)

$$= (-1)^{p(i)p(j)}D_i(Q_j)(q_i)\cdot X_j(-1)^{p(i)p(X_j)}$$

$$D_i(Q_j)(q_i)\cdot X_j = X_jD_i(Q_j)(q_i)(-1)^{p(X_j)p(Q_j)}.$$

Since

$$p(i)p(R_i) + p(X_j)[p(Q_i)+p(j)] + p(i)p(j) + p(i)p(X_j) + p(X_j)p(Q_j) =$$
$$= p(i)[p(Q_j)+1] \pmod 2,$$

collecting together terms in (2.71) we can replace the second term in (2.70) by

$$\sim \sum X_j\int_0^1 dtD_i(Q_j)(q_i)(-1)^{p(i)[p(Q_j)+1]} =$$
$$= \sum X_i\int_0^1 dtD_j(A_t(R_i))(q_j)(-1)^{p(j)[p(R_i)+1]}.$$

Thus, (2.70) becomes

$$X(H) \sim \sum_i X_i\int_0^1 dt\{A_t(R_i) + \sum_j D_j(A_t(R_i))(q_j)(-1)^{p(j)[p(R_i)+1]}\}. \tag{2.72}$$

Now, from the obvious formula

$$\frac{\partial}{\partial q_j^{(g|\nu)}}A_t = t\,A_t\frac{\partial}{\partial q_j^{(g|\nu)}},$$

we conclude that

$$D_j(A_t(R_i))(q_j) = \sum\frac{\partial}{\partial q_j^{(g|\nu)}}[A_t(R_i)]\cdot\widehat{g}\partial^\nu(q_i) =$$

$$= \sum tA_t\left(\frac{\partial R_i}{\partial q_j^{(g|\nu)}}\right)\cdot q_j^{(g|\nu)} = A_t\left(\sum\frac{\partial R_i}{\partial q_j^{(g|\nu)}}q_j^{(g|\nu)}\right),$$

and hence (2.72) becomes

$$X(H) \sim \sum_i X_i \int_0^1 dt\, A_t \left[R_i + \sum q_j^{(g|\nu)} \frac{\partial R_i}{\partial q_j^{(g|\nu)}} \right]. \tag{2.73}$$

Since the Euler theorem on degree-homogeneous functions, obviously, holds true also in the presence of odd variables, we have

$$\sum q_j^{g|\nu)} \frac{\partial R_i}{\partial q_j^{(g|\nu)}} = deg(R_i) \cdot R_i,$$

and therefore, (2.73) is transformed into

$$X(H) \sim \sum X_i \int_0^1 dt\, A_t(R_i)[1 + deg\,(R_i)] =$$

$$= \sum X_i \int_0^1 dt\, t^{deg\,(R_i)}\, [1 + deg(R_i)]\, R_i = \sum X_i\, R_i. \qquad \blacksquare$$

2.74. Remark. When $N = |I| = \infty$ and $\overline{R}$ is *not* a finite vector, Theorem 2.67 fails unless appropriate growth conditions are imposed on R_i's (and the basic philosophy of the calculus of variations is redesigned as well).

2.75. Remark. The same method as the one used to prove Theorem 2.67, can be easily adjusted to prove the relation

$$Ker\,\delta = Im\mathcal{D} + K \tag{2.76}$$

of Remark 2.14. Indeed, suppose $\delta(H) = 0$. Let $X \in D^{ev}$ be given as

$$X(q_i) = q_i \to X = \sum q_i^{(g|\nu)} \frac{\partial}{\partial q_i^{(g|\nu)}},$$

so that

$$\frac{\partial}{\partial t} A_t = t^{-1} A_t\, X, \tag{2.77}$$

with A_t taken from the Proof of Theorem 2.67. Since

$$\int_0^1 dt\, t^{-1} A_t(f) = \frac{1}{deg(f)} f, \; deg\,(f) > 0, \tag{2.78}$$

we have

$$H - H|_{q=0} = \int_0^1 \frac{\partial}{\partial t}(A_t(H))dt = \int_0^1 dtt^{-1} A_t X(H) = \frac{1}{deg(H)} X(H),$$

so that

$$H = H|_{q=0} + \frac{1}{deg(H)} X(H), \quad H|_{q=0} = A_0(H), \tag{2.79}$$

and $X(H) \sim < X, \delta(H) >= 0$ by (2.17). ∎

§3. SuperHamiltonian Formalism and Infinite-Dimensional Stable Lie Superalgebras

The main objects in this section are: superHamiltonian structures; canonical maps between superHamiltonian structures; stable Lie superalgebras over commutative superalgebras with calculus; generalized two-cocycles on stable Lie superalgebras. The main results are: an algebraic criterion of the superHamiltonian property; a large class of superHamiltonian structures: even superskewsymmetric operators with coefficients in the basic commutative superalgebra K; a criterion for a map to be canonical; a one-to-one correspondence between linear Hamiltonian structures and stable Lie superalgebras, and between affine Hamiltonian structures and generalized two-cocycles on stable Lie superalgebras.

Let k, K, and $C = K[q_i^{(g|\nu)}]$ be as in §2.

3.1. Definition. An even k–linear map $\Gamma : C \to D^{ev}(C)$ is called *superHamiltonian* if the following conditions are satisfied:

(3.1*i*) $\{H, F\} \sim -(-1)^{p(H)p(F)}\{F, H\}, \quad \forall H, F \in C,$

where $\{H, F\} = X_H(F)$ is called the Poisson bracket, and $X_H = \Gamma(H)$;

(3.1*ii*) $X_{\{H,F\}} = [X_H, X_F], \quad \forall H, F \in C,$

where the commutator on the right is understood in the Lie superalgebra sense: $[a, b] = ab - (-1)^{p(a)p(b)}ba$;

(3.1*iii*) There exist two operators $B^0, B^1 : C^N \to C^N, N = |I|$, such that

(3.2) $\overline{X}_H = B^{p(H)}\left(\frac{\delta H}{\delta \overline{q}}\right),$

where $\overline{X}_H = X_H(\overline{q})$ by (2.15).

$(3.1i^4)$ The properties $(3.1i - iii)$ remain true for any (differential-difference) extension $K' \supset K$ over k, i.e., for any commutative superalgebra extension K' on which the action of G and ∂'s is compatible with their action on K.

3.3. Remark. As usual in working with superobjects, we give definitions and prove formulae for $\mathbf{Z}_2$-homogeneous elements only, and then extend definitions and formulae by additivity to all elements (see [Ka 2; Le]).

3.4. Remark. The reader with roots in classical mechanics may wonder what has happened with two expected requirements on the Poisson bracket: the derivation property

$$(3.5) \quad \{H, FR\} \sim \{H, F\}R + \{H, R\}F(-1)^{p(F)p(R)}, \quad \forall H, F, R \in C;$$

and the graded Jacobi identity

$$(3.6) \quad \{H, \{F, R\}\} \sim \{\{H, F\}R\} + (-1)^{p(H)p(F)}\{F, \{H, R\}\}, \forall H, F, R, \in C.$$

The property (3.5) is already meaningless outside the very degenerate area of classical mechanics (i.e., the case when G and ∂'s are absent) even (e.g., in models of classical field theory) when B^0 and B^1 are homomorphisms of right C-modules, the ultimate reason being that ImD is *not* a C-submodule in C. (Incidentally, the calculus of variations can be thought of as an apparatus providing a nontrivial module structure in $\Omega^1(C)/ImD$ even though ImD is not a submodule.) With respect to the graded Jacobi identity (3.6), it is interchangeable with the basic Hamiltonian property $(3.1ii)$ but *only* when the stability condition $(3.1i^4)$ is invoked (see Theorem 3.71 below). We will not need this fact for awhile. Suffice it to notice that (3.6) follows directly from $(3.1ii)$ by applying both parts of the equality $(3.1ii)$ to R.

3.7. Remark. The only *really new* feature of the Hamiltonian formalism in the presence of Grassmann-type variables is the property (3.1*iii*): one has now *two* defining matrices instead of just *one* in the purely commutative case.

3.8. Lemma. Denote $B = B^0$. Then

$$(3.9)\quad B^{s\dagger} = -B,$$

$$(3.10)\quad B^1_{ij} = (-1)^{p(i)} B_{ij}.$$

Proof. (a) For arbitrary $H, F \in C_0$, we have from (3.1*i*), (3.2), and (2.17):

$$\left[B\left(\frac{\delta H}{\delta \overline{q}}\right)\right]^t \frac{\delta F}{\delta \overline{q}} =< X_H, \delta(F) >\sim X_H(F) \sim -X_F(H) \sim - < X_F, \delta(H) >=$$

$$= -\left[B\left(\frac{\delta F}{\delta \overline{q}}\right)\right]^t \left(\frac{\delta H}{\delta \overline{q}}\right) \text{ [by Proposition 2.44] } \sim -\left[B^{s\dagger}\left(\frac{\delta H}{\delta \overline{q}}\right)\right]^t \frac{\delta F}{\delta \overline{q}}.$$

(3.11)

Therefore

$$(3.12)\quad \left[(B + B^{s\dagger})\left(\frac{\delta H}{\delta \overline{q}}\right)\right]^t \frac{\delta F}{\delta \overline{q}} \sim 0, \quad \forall H, F \in C_0.$$

Now set $K' = K[Q_i^{(g|\nu)}], i \in I$, with $p(Q_i) = p(q_i)$, and take $F = q_i Q_i$ (no sum on i). Then $\dfrac{\delta F}{\delta q_j} = \delta_j^i Q_i$, and denoting $\overline{Y} = (B + B^{s\dagger})\left(\dfrac{\delta H}{\delta \overline{q}}\right)$, we have from (3.12) $Y_i Q_i \sim 0$ (no sum on i). Hence, $0 = \dfrac{\delta}{\delta Q_i}(Y_i Q_i) = (-1)^{p(i)} Y_i$, so that

$$(3.13)\quad \left(B + B^{s\dagger}\right)\left(\frac{\delta H}{\delta \overline{q}}\right) = 0, \quad \forall H \in C_0.$$

Now take H in (3.13) to be $q_i Q_i$ (no sum on i). Then

$$(3.14)\quad (B + B^{s\dagger})_{ji}(Q_i) = 0, \qquad \forall j, i \in I.$$

From Lemma 3.19 below it follows that $B + B^{s\dagger} = 0$, which is (3.9).

(b) Now take $H \in C_0, F \in C_1$. Analogously to (3.11), we have

$$\left[B\left(\frac{\delta H}{\delta \overline{q}}\right)\right]^t \frac{\delta F}{\delta \overline{q}} = < X_H, \delta(F) > \sim X_H(F) \sim -X_F(H) \sim - < X_F, \delta(H) > = $$
$$= - \left[B^1\left(\frac{\delta F}{\delta \overline{q}}\right)\right]^t \frac{\delta H}{\delta \overline{q}}.$$

Hence,

$$-\sum B^1_{ij}\left(\frac{\delta F}{\delta q_j}\right)\cdot\frac{\delta H}{\delta q_i} = -\left[B^1\left(\frac{\delta F}{\delta \overline{q}}\right)\right]^t \frac{\delta H}{\delta \overline{q}} \sim \left[B\left(\frac{\delta H}{\delta \overline{q}}\right)\right]^t \frac{\delta F}{\delta \overline{q}} =$$

(3.15)

$$= \sum B_{ji}\left(\frac{\delta H}{\delta q_i}\right)\cdot\frac{\delta F}{\delta q_j}\ \text{[by (2.37)]} \sim \sum B^{\dagger}_{ji}\left(\frac{\delta F}{\delta q_j}\right)\cdot\frac{\delta H}{\delta q_i}(-1)^{p(i)[p(j)+1]}.$$

Since $B^{s\dagger} = -B$ by (3.9), we can use (2.43) to transform (3.15) into

$$(3.16)\qquad \sum\left[B_{ij} - (-1)^{p(i)} B^1_{ij}\right]\left(\frac{\delta F}{\delta q_j}\right)\cdot\frac{\delta H}{\delta q_i} \sim 0, \quad \forall H \in C_0, \forall F \in C_1.$$

Taking $H = q_i Q_i$ and following the same route as in deriving (3.13), we conclude that

$$(3.17)\quad \sum_j\left[B_{ij} - (-1)^{p(i)} B^1_{ij}\right]\left(\frac{\delta F}{\delta q_j}\right) = 0.$$

Taking now $F = q_j R_j$ (no sum on j), with $K' = K[R_i^{(g|\nu)}], i \in I, p(R_i) = p(q_i) + 1$, we find from (3.17) that

$$(3.18)\quad \left(B_{ij} - (-1)^{p(i)} B^1_{ij}\right)(R_j) = 0 \text{ (no sum on } j),$$

and again from Lemma 3.19 below it follows that $B_{ij} - (-1)^{p(i)} B^1_{ij} = 0$, which is (3.10). ▮

3.19. Lemma. Let $A : C \to C$ be an operator. Let $K' = K[Q_s^{(g|\nu)}], s \in S, C' = K'[q_i^{(g|\nu)}], i \in I$. Extend A naturally to act on C'. If $A(Q_1) = 0$ then $A = 0$.

Proof. Let $A = \sum A^{g|\nu} \hat{g} \partial^\nu, A^{g|\nu} \in C$. Then $A(Q_1) = \sum A^{g|\nu} Q_1^{(g|\nu)}$. Since $Q_1^{(g|\nu)}$'s are independent variables in C' and $A^{g|\nu}$'s do not involve Q's, we conclude that $A^{g|\nu} = 0$. ▮

3.20. Corollary.

$$(3.21)\quad (\overline{X}_H)_i = (-1)^{p(i)p(H)} \sum_j B_{ij} \left(\frac{\delta H}{\delta q_j}\right),$$

$$(3.22)\quad \{H, F\} \sim \sum (-1)^{p(i)p(H)} B_{ij} \left(\frac{\delta H}{\delta q_j}\right) \cdot \frac{\delta F}{\delta q_i}.$$

Proof follows from (3.2) and (3.10). ▮

3.23. Remark. From Lemma 3.8 it follows that one can work with only one matrix, namely $B = B^0$, instead of both B^0 and B^1. If we write B in the block form as $B = \begin{pmatrix} \alpha & \beta \\ \gamma & \rho \end{pmatrix}$ where α is a $N_0 \times N_0-, \beta$ is a $N_0 \times N_1-, \gamma$ is a $N_1 \times N_0-$, and ρ is a $N_1 \times N_1-$ matrix, $N_0 = |I_0|, N_1 = |I_1|$, then (3.9) means that $\alpha^\dagger = -\alpha, \rho^\dagger = \rho, \beta^\dagger = -\gamma$, with "†" being the commutative adjoint defined by (2.41).

3.24. Remark. To make sure that the relations (3.9), (3.10) together are *equivalent* to (3.1i) we need also to verify that (3.1i) is satisfied when both H and F are odd. To check this, we use (3.22):

$$\{H, F\} \sim \sum (-1)^{p(i)} B_{ij} \left(\frac{\delta H}{\delta q_j}\right) \cdot \frac{\delta F}{\delta q_i} \sim$$

$$\sim \sum (-1)^{p(i)} (B_{ij})^\dagger \left(\frac{\delta F}{\delta q_i}\right) \cdot \frac{\delta H}{\delta q_j} \quad (-1)^{[p(i)+1][p(j)+1]} [\text{by } (3.9), (2.43)] =$$

$$= \sum (-1)^{p(i)+p(i)p(j)+1} B_{ji} \left(\frac{\delta F}{\delta q_i}\right) \cdot \frac{\delta H}{\delta q_j} (-1)^{p(i)[p(j)+1]+p(j)+1} =$$

$$= \sum (-1)^{p(j)} B_{ji} \left(\frac{\delta F}{\delta q_i}\right) \cdot \frac{\delta H}{\delta q_j} \quad [\text{by } (3.22)] \quad \sim \{F, H\}.$$

From now on we assume that B is even superskewsymmetric. Our main goal is to find necessary and sufficient conditions for a given superskewsymmetric matrix B to be superHamiltonian, i.e., to define a superHamiltonian structure which means, in turn, that the formula $(3.1ii)$ is stably satisfied. We shall achieve this by subsequently transforming the nonoperator equality $(3.1ii)$ into an operator one. It will take some preparation.

3.25. Lemma. The equality $(3.1ii)$ is stably satisfied for all H, F if it is stably satisfied for all even H, F.

Proof. Set $K' = K[\theta]$, with $p(\theta) = 1, \widehat{g}(\theta) = \theta, \forall g \in G, \partial_s(\theta) = 0, s = 1, ..., m$. We have,

$$(3.26) \quad X_{\theta H} = \theta X_H, \qquad \forall H \in C.$$

Indeed, by (3.2) and (3.10) we have

$$(\overline{X}_{\theta H})_i = \sum B_{ij}^{p(H)+1} \left(\frac{\delta(\theta H)}{\delta q_j}\right) = \sum B_{ij}^{p(H)+1} \theta (-1)^{p(j)} \left(\frac{\delta H}{\delta q_j}\right) =$$

$$= \theta \sum (-1)^{p(i)+p(j)} B_{ij}^{p(H)+1} (-1)^{p(j)} \left(\frac{\delta H}{\delta q_j}\right) =$$

$$= \theta \sum B_{ij}^{p(H)} \left(\frac{\delta H}{\delta q_j}\right) = \theta (\overline{X}_H)_i,$$

where we used the equality $p(B_{ij}^{\alpha}) = p(i) + p(j), \forall \alpha \in \mathbf{Z}_2$ (since Γ is even), and the fact that θ is an odd constant as far as G, ∂'s and δ are concerned.

Assume now that (3.1*ii*) is satisfied for all even H and F. Suppose that H is odd. Then θH is even and, by (3.26),

$$\{\theta H, F\} = X_{\theta H}(F) = \theta X_H(F) = \theta\{H, F\},$$

so that, again by (3.26), the left-hand-side of (3.1*ii*) becomes

$$X_{\{\theta H,F\}} = \theta X_{\{H,F\}}. \tag{3.27}$$

On the other hand, the right-hand-side of (3.1*ii*) becomes

$$\begin{aligned}[X_{\theta H}, X_F] &= X_{\theta H} X_F - (-1)^{p(\theta H)p(F)} X_F X_{\theta H} = \\ &= \theta X_H X_F - X_F \theta X_H = \theta[X_H X_F - (-1)^{p(F)} X_F X_H] = \\ &= \theta[X_H X_F - (-1)^{p(F)p(H)} X_F X_H] = \theta[X_H, X_F].\end{aligned} \tag{3.28}$$

(a) If F is even, we can use (3.1*ii*) for even θH and F. Equating (3.27) with (3.28) (as maps of C into θC) and dividing out the resulting equality by θ, we arrive at (3.1*ii*) with odd H and even F. (b) If now F is also odd then θF is even, so by (a)

$$X_{\{H,\theta F\}} = [X_H, X_{\theta F}].$$

But

$$\{H, \theta F\} \sim -\{\theta F, H\} = -\theta\{F, H\},$$

So that

$$X_{\{H,\theta F\}} = X_{-\theta\{F,H\}} = -\theta X_{\{F,H\}}.$$

On the other hand,

$$[X_H, X_{\theta F}] = -[X_{\theta F}, X_H] \text{ [by (3.28)] } = -\theta [X_H, X_F],$$

So that $(3.1ii)$ results again.

Finally, none of the arguments we have employed had made any use of the specific properties of the superalgebra K. This means that changing K into any of its extensions would not affect the arguments, and therefore it would not affect the result, which is exactly the meaning of stability. ∎

3.29. Remark. We shall, as a rule, omit the stability reasoning demonstrated above whenever no specific information about K is used.

From now on, until the appearance of Lie superalgebras, in working with the relation $(3.1ii)$ we can, and shall, consider only even H and F, thanks to Lemma 3.25.

3.30. Lemma. For even H and F, the relation $(3.1ii)$ is equivalent to

$$(3.31)\quad B\frac{\delta}{\delta\bar{q}}\left\{\left[B\left(\frac{\delta H}{\delta\bar{q}}\right)\right]^t\frac{\delta F}{\delta\bar{q}}\right\} = D^0\left[B\left(\frac{\delta F}{\delta\bar{q}}\right)\right]B\left(\frac{\delta H}{\delta\bar{q}}\right) - \\ -D^0\left[B\left(\frac{\delta H}{\delta\bar{q}}\right)\right]B\left(\frac{\delta F}{\delta\bar{q}}\right),$$

where the even Fréchet derivative D^0 is defined by (2.21).

Proof. Two evolution fields coincide if and only if they yield the same result acting on the vector $\bar{q} = (q_i)$. Applying each side of the equality $(3.1ii)$ to $\bar{q}$ we obtain, using $(3.1iii)$,

$$X_{\{H,F\}}(\bar{q}) = B\left(\frac{\delta\{H,F\}}{\delta\bar{q}}\right) \quad \text{[by (3.22) and Corollary 2.7]}$$

$$= B\frac{\delta}{\delta\bar{q}}\left\{\left[B\left(\frac{\delta H}{\delta\bar{q}}\right)\right]^t\frac{\delta F}{\delta\bar{q}}\right\},$$

$[X_H, X_F](\bar{q}) = X_H(\overline{X}_F) - X_F(\overline{X}_H)$ [by Corollary 2.31] $= D^0(\overline{X}_F)(\overline{X}_H) -$

$$-D^0(\overline{X}_H)(\overline{X}_F) = D^0\left[B\left(\frac{\delta F}{\delta\bar{q}}\right)\right]B\left(\frac{\delta H}{\delta\bar{q}}\right) - D^0[B\left(\frac{\delta H}{\delta\bar{q}}\right)]B\left(\frac{\delta F}{\delta\bar{q}}\right).$$ ∎

3.32. Definition. For a column vector $R \in C^N$, we denote by R^{st} its super-transpose which is a row vector with the components

(3.33) $(R^{st})_i = (-1)^{p(i)}R_i$,

reserving the notation R^t for the usual transpose.

3.34. Proposition. If R and S are even vectors in C^N then

(3.35) $R^tS = S^{st}R$.

Proof. We have,

$R^tS = \sum R_iS_i = \sum(-1)^{p(i)p(i)}S_iR_i = S^{st}R$. ∎

3.36. Lemma. If $X \in D^{ev}, S \in C^N$, and $H \in C$ are all even then

(3.37) $$\left[X\left(\frac{\delta H}{\delta\bar{q}}\right)\right]^{st} S \sim \overline{X}^t\left[D^0\left(\frac{\delta H}{\delta\bar{q}}\right)(S)\right].$$

Proof. Denote $R = \dfrac{\delta H}{\delta\bar{q}}$. Then R is even, $X(R)$ is even, and hence

$$\left[X\left(\frac{\delta H}{\delta\bar{q}}\right)\right]^{st} S =$$

$= \sum(-1)^{p(i)}X(R_i)S_i$ [by Lemma 2.24, (2.21), and (2.19)]

$= \sum(-1)^{p(i)}(-1)^{p(j)[p(i)+1]}D_j(R_i)(X_j)\cdot S_i$[by (2.37)] $\sim$

$$\sim \sum (-1)^{p(i)+p(j)p(i)+p(j)} [D_j(R_i)]^\dagger (S_i) \cdot X_j (-1)^{p(i)p(j)} \text{ [by (2.66)]} =$$
$$= \sum (-1)^{p(i)+p(j)+p(i)p(j)} D_i(R_j)(S_i) \cdot X_j =$$
$$= \sum (-1)^{p(i)+p(j)+p(i)p(j)} X_j \, D_i(R_j)(S_i)(-1)^{p(j)p(j)} \text{ [by (2.29)]} =$$
$$= \sum X_j [D^0(R)]_{ji}(S_i) = \overline{X}^t [D^0(R)(S)]. \quad ■$$

3.38. Definition. For an even evolution field $X \in D^{ev}$ and an operator $A : C \to C$, the action of X on A is defined as $X(\sum A^{g|\nu} \widehat{g} \partial^\nu) = \sum X(A^{g|\nu}) \widehat{g} \partial^\nu$. If A is a matrix operator then X acts on A matrix elements-wise.

3.39. Proposition. If $X \in D^{ev}$ is even, $A : C^N \to C^N$ is an operator, and $R \in C^N$ is an even vector, then

$$(3.40) \quad X(A)(R) = XA(R) - AX(R) =: [X, A](R) =$$
$$= ([D^0, A](R))(\overline{X}) := [D^0(AR) - AD^0(R)](\overline{X}).$$

Proof. Since X acts on A matrix elements-wise, it is enough to check the first equality in (3.40) for A and R being scalars. In this case, if $A = \sum A^{g|\nu} \widehat{g} \partial^\nu$ then

$$(3.41) \qquad X(A)(R) = \left[X(\sum A^{g|\nu} \widehat{g} \partial^\nu)\right](R) = \left(\sum X(A^{g|\nu}) \widehat{g} \partial^\nu\right)(R),$$

$[X, A](R) = X[\sum A^{g|\nu} \widehat{g} \partial^\nu (R)] - \sum A^{g|\nu} \widehat{g} \partial^\nu X(R)$ [since X commutes with $\widehat{g} \partial^\nu$] $= \sum X(A^{g|\nu}) \widehat{g} \partial^\nu (R) + \sum A^{g|\nu} \widehat{g} \partial^\nu X(R) - \sum A^{g|\nu} \widehat{g} \partial^\nu X(R) =$

$= \sum X(A^{g|\nu}) \widehat{g} \partial^\nu (R)$,

which is the same as (3.41). Now, using Corollary 2.31 we obtain $X(A)(R) = X(AR) - AX(R) = D^0(AR)(\overline{X}) - AD^0(R)(\overline{X})$. ■

3.42. Corollary. With respect to each of the vectors R and $\overline{X}$, the expression $([D^0, A](R))(\overline{X})$ is an operator.

Proof. Indeed, by (3.40) this expression equals to $X(A)(R)$ and, thus, involves only operations of the type $\{\sum \varphi \hat{g}\, \partial^{\nu} | \varphi \in C\}$ applied to the components of both R and $\overline{X}$. ∎

We are now in a position to derive the main technical result of the superHamiltonian formalism (which is a generalization of Lemma 1.4 in [K – M]).

3.43. Theorem. For an even matrix operator $B : C^N \to C^N$ and even vectors $R, S \in C^N$, denote by $< B, R, S >$ a column vector defined as

$$< B, R, S >_i = (-1)^{p(i)} \left(([D^0, B](R))^{s\dagger}(S) \right)_i . \tag{3.44}$$

If B is even superskewsymmetric then for any even H, $F \in C$,

$$\frac{\delta}{\delta \overline{q}} \left[\left(B \left(\frac{\delta H}{\delta \overline{q}} \right) \right)^t \frac{\delta F}{\delta \overline{q}} \right] = D^0 \left(\frac{\delta F}{\delta \overline{q}} \right) B \left(\frac{\delta H}{\delta \overline{q}} \right) - D^0 \left(\frac{\delta H}{\delta \overline{q}} \right) B \left(\frac{\delta F}{\delta \overline{q}} \right) + $$

$$+ < B, \frac{\delta H}{\delta \overline{q}}, \frac{\delta F}{\delta \overline{q}} > . \tag{3.45}$$

Proof. To prove (3.45) we will show that for any even $X \in D^{ev}$, the product of $\overline{X}^t$ with the left-hand-side of (3.45) differs by $Im D$ from the product of $\overline{X}^t$ with the right-hand-side of (3.45), and then appeal to (2.17). We have

$$\overline{X}^t \frac{\delta}{\delta \overline{q}} \left[\left(B \left(\frac{\delta H}{\delta \overline{q}} \right) \right)^t \frac{\delta F}{\delta \overline{q}} \right] \quad \text{[by (2.17)]} \quad \sim X \left[\left(B \left(\frac{\delta H}{\delta \overline{q}} \right) \right)^t \frac{\delta F}{\delta \overline{q}} \right] =$$

(3.46)

$$= \left[X(B) \left(\frac{\delta H}{\delta \overline{q}} \right) \right]^t \frac{\delta F}{\delta \overline{q}} + \left[BX \left(\frac{\delta H}{\delta \overline{q}} \right) \right]^t \frac{\delta F}{\delta \overline{q}} + \left[B \left(\frac{\delta H}{\delta \overline{q}} \right) \right]^t X \left(\frac{\delta F}{\delta \overline{q}} \right)$$

We transform separately each of the three summands in (3.46), denoting $R = \frac{\delta H}{\delta \overline{q}}$ and $S = \frac{\delta F}{\delta \overline{q}}$:

1) $[X(B)(R)]^t S$ [by (3.40)] $= \{([D^0, B](R))(\overline{X})\}^t S$ [by (2.45)] $\sim$

$\sim \left\{([D^0, B](R))^{s\dagger}(S)\right\}^t \overline{X}$ [by Proposition 3.34] $=$

$= \overline{X}^{st} \left\{([D^0, B](R))^{s\dagger}(S)\right\}$ [by (3.33) and (3.44)]$=$

(3.47) $= \overline{X}^t < B, R, S >$;

2) $[BX(R)]^t S$ [since $B^{s\dagger} = -B$] $\sim -[B(S)]^t X(R)$ [by (3.35)] $=$

(3.48) $= -X(R)^{st} B(S)$ [by (3.37)] $\sim -\overline{X}^t [D^0(R)(B(S))]$;

3) $[B(R)]^t X(S)$ [by (3.35)] $= X(S)^{st} B(R)$ [by (3.37)] $\sim$

(3.49) $\sim \overline{X}^t [D^0(S)(B(R))]$.

Substituting (3.47)–(3.49) into (3.46) we get (3.45). ∎

Substituting now (3.45) into (3.31) we immediately obtain

3.50. Lemma. For even H and F, the relation (3.1*ii*) is equivalent to

(3.51) $$B < B, \frac{\delta H}{\delta \overline{q}}, \frac{\delta F}{\delta \overline{q}} > = \left([D^0, B]\left(\frac{\delta F}{\delta \overline{q}}\right)\right) B \left(\frac{\delta H}{\delta \overline{q}}\right) - \left([D^0, B]\left(\frac{\delta H}{\delta \overline{q}}\right)\right) B \left(\frac{\delta F}{\delta \overline{q}}\right) .$$

Now we can derive the main result of the superHamiltonian formalism.

3.52. Theorem. An even superskewsymmetric matrix B is superHamiltonian iff (3.1*ii*) is stably satisfied for arbitrary even *linear* functions, i.e. for any even H and F of the form $H = \sum q_i X_i$, $F = \sum q_j Y_j$, with X_i's and Y_j's taken from arbitrary extension $K' \supset K$.

Proof. If $H = \sum q_i X_i$ and $F = \sum q_j Y_j$ are even and linear then the vectors $\frac{\delta H}{\delta \bar{q}} = X = (X_i)$ and $\frac{\delta F}{\delta \bar{q}} = Y = (Y_i)$ are even, and (3.51) becomes

$$(3.53) \quad B < B, X, Y >= ([D^0, B](Y))\, B(X) - ([D^0, B](X))B(Y),$$

which we assume is satisfied for any even $X, Y \in K'^N$. Now, by (3.44) and Corollary 3.42, each side of (3.53) is a bilinear operator acting on components of X and Y. Fixing X (or Y) we obtain an equality involving two operators (on each side) acting on arbitrary $Y \in K'^N$ (or $X \in K'^N$). By Lemma 3.19, this implies that we have in fact an operator identity and, thus, (3.53) is valid for arbitrary even $X, Y \in C'^N, C' = K'[q_i^{(g|\nu)}]$. In particular, (3.53) is valid for $X = \frac{\delta H}{\delta \bar{q}}, Y = \frac{\delta F}{\delta \bar{q}}$ with arbitrary even $H, F \in C'$. Hence, (3.51) is satisfied.∎

3.54. Corollary. For a given even superskewsymmetric matrix B, to check the superHamiltonian property of B it is necessary and sufficient to check the following identity

$$(3.55) \quad B \frac{\delta}{\delta \bar{q}} [B(X)^t Y] = D^0(BY)B(X) - D^0(BX)B(Y)$$

in the superalgebra $K'[q_i^{(g|\nu)}]$, where $K' = K[X_i^{(g|\nu)}, Y_i^{(g|\nu)}], i \in I, p(X_i) = p(Y_i) = p(i)$.

Proof. By Theorem 3.52, to check the superHamiltonian property of B one can work with even linear functions only. For such functions, the superHamiltonian condition (3.1*ii*) in the form (3.31) becomes (3.55). To check (3.55) for arbitrary $K' \supset K$ it is enough, by Lemma 3.19, to check it for the universal case $K' = K[X_i^{(g|\nu)}]$ when $X_i^{(g|\nu)}$'s and $Y_i^{(g|\nu)}$'s are considered as independent generators. ∎

We can now describe a large class of superHamiltonian structures.

3.56. Definition. We say that an operator A is with coefficients in (a commutative superalgebra) $K' \supset K$ if all matrix elements of A are of the form $\{\sum \varphi \hat{g} \partial^\nu | \varphi \in K'\}$.

3.57. Theorem. If B is an even superskewsymmetric matrix with coefficients in K then B is superHamiltonian.

Proof. By Lemma 3.58 below, $[D^0, B](R) = 0$ for any even $R \in C'^N$.

Thus, (3.51) becomes $0 = 0$ since, by (3.44), $< B, \frac{\delta H}{\delta \overline{q}}, \frac{\delta F}{\delta \overline{q}} >$ vanishes when $[D^0, B]\left(\frac{\delta H}{\delta \overline{q}}\right) = 0$. ∎

3.58. Lemma. Let $A : C^N \to C^N$ be an even operator with coefficients in K. Then $[D^0, A](R) = 0$ for any even $R \in C'^N$.

Proof. The operator $[D^0, A](R)$ vanishes if $([D^0, A](R)(\overline{X}) = 0$ for any even $\overline{X} \in C'^N$ (Lemma 3.19). By (3.40), for any even $X \in D^{ev}(C')$,

$$X(A)(R) = 0 = XA(R) - AX(R) \text{ [by (2.32)] } =$$

$$= (D^0(A(R)) - AD^0(R))(\overline{X}) = [D^0, A](R))(\overline{X}).$$ ∎

3.59. Remark. If $\mathcal{G}$ is a finite-dimensional Lie algebra over a field then the ring of polynomial functions on the dual space $\mathcal{G}^*$ to $\mathcal{G}$ possesses a natural Hamiltonian structure whose associated Poisson bracket can be defined very simply by: (a) being a derivation with respect to each argument; and (b) coinciding with the commutator in $\mathcal{G}$ for *linear functions* on $\mathcal{G}^*$ considered as elements of $\mathcal{G}$. Theorem 3.52 may be thought of as a nonlinear infinite-dimensional generalization of this definition.

We now consider the problem of canonical maps. Let $C_1 = K[u_j^{(g|\nu)}]$, $j \in J = J_0 \cup J_1$, be another commutative superalgebra. Suppose matrices B and B_1 define superHamiltonian structures in C and C_1 respectively. Let $\Phi : C \to C_1$ be a homomorphism (see §2).

3.60. Definition. Φ is (stably) canonical if evolution fields X_H (in C') and $X_{\Phi(H)}$ (in C'_1) are compatible for any $H \in C' = K'[q_i^{(g|\nu)}]$, for any $K' \supset K$:

$$\Phi X_H = X_{\Phi(H)} \Phi . \tag{3.61}$$

3.62. Lemma. To check the canonical property of Φ it is enough to consider only even H's in (3.61).

Proof. Suppose (3.61) is satisfied for even H's. Let $F \in C'$ be odd. As in the Proof of Lemma 3.25, set $K'' = K'[\theta]$. Then θF is even, and $\Phi(\theta F) = \theta\Phi(F)$. Using (3.61) for $H = \theta F$ we obtain, with the help of (3.26):

$$\begin{aligned} \theta\Phi X_F = \Phi\theta X_F = \Phi X_{\theta F} = X_{\Phi(\theta F)}\Phi = \\ = X_{\theta\Phi(F)}\Phi = \theta X_{\Phi(F)}\Phi, \end{aligned}$$

and dividing out by θ we see that (3.61) is satisfied for odd H's as well. ∎

3.63. Theorem. A homomorphism Φ is canonical if and only if

$$\Phi(B) = D^0(\overline{\Phi})B_1 D(\overline{\Phi})^\dagger , \tag{3.64}$$

where $\overline{\Phi} = (\Phi_i = \Phi(q_i))$ and Φ acts on operators matrix elements-wise by the rule : $\Phi(\sum f\hat{g}\partial^\nu) = \sum \Phi(f)\hat{g}\partial^\nu$.

Proof. We work out (3.61) for even H's. Since Φ, X_H, and $X_{\Phi(H)}$ all commute with the action of G, ∂'s, and (arbitrary) K', (3.61) is equivalent to $\Phi X_H(\overline{q}) = X_{\Phi(H)}\Phi(\overline{q})$, or $\Phi(\overline{X}_H) = X_{\Phi(H)}(\overline{\Phi})$. In other words,

$$0 = \Phi(\overline{X}_H) - X_{\Phi(H)}(\overline{\Phi}) \quad \text{[by (3.21), (2.32)]} =$$

$$= \Phi\left(B\left(\frac{\delta H}{\delta \overline{q}}\right)\right) - D^0(\overline{\Phi})B_1\left(\frac{\delta(\Phi(H))}{\delta \overline{u}}\right) \text{[by (2.53)]} = \tag{3.65}$$

$$= [\Phi(B) - D^0(\overline{\Phi})B_1 D(\overline{\Phi})^\dagger]\Phi\left(\frac{\delta H}{\delta \overline{q}}\right).$$

Thus, if (3.64) is satisfied then Φ is canonical. Conversely, take H to be linear: $H = \sum q_j X_j, X_j \in K'$. Then $\Phi\left(\frac{\delta H}{\delta \overline{q}}\right) = \overline{X} \in K'^N$, and by Lemma 3.19 we obtain (3.64) from (3.65). ∎

3.66. Remark. Applying (3.61) to arbitrary $F \in C$, we can rewrite (3.61) in an equivalent form

$$\Phi(\{H, F\}) = \{\Phi(H), \Phi(F)\}, \quad \forall H,\ F \in C. \tag{3.67}$$

In practice, one often defines a (stable) canonical map as an even homomorphism $\Phi : C \to C_1$ which preserves the Poisson brackets module ImD:

$$\Phi(\{H, F\}) \sim \{\Phi(H), \Phi(F)\}, \quad \forall H, F \in C'. \tag{3.68}$$

Let us show that both definitions are equivalent. First, we rewrite (3.68) in the form

$$[\Phi X_H - X_{\Phi(H)}\Phi](C') \sim 0 \quad \text{in}\, C_1', \quad \forall H \in C. \tag{3.69}$$

Let us check that if H is even in (3.69) then (3.69) implies (3.61) (for this H). By Lemma 3.62, it follows that Φ is canonical in the first sense.

Denote $Z = \Phi X_H - X_{\Phi(H)}\Phi$. Z is an even (quasi-evolution) derivation of C' into C'_1 along Φ, which commutes with the action of G, ∂'s and K'; also, $Z(C') \sim 0$. From Lemma (3.70) below it follows that $Z = 0$. ∎

3.70. Lemma. Let $\Phi : C \to C_1$ be a homomorphism, and $Z : C \to C_1$ be a derivation along Φ over K. If $Z(fk') \sim 0$ for a fixed $f \in C$ and $\forall k'$ from arbitrary extension $K' \supset K$ then $Z(f) = 0$. In particular, if $Z(C) \sim 0, \forall K' \supset K$ then $Z = 0$.

Proof. Take $K' = K[Q^{(g|\nu)}], p(Q) = 0$. For $f \in C, 0 \sim Z(fQ) = Z(f)Q$, hence $0 = \dfrac{\delta}{\delta Q}[Z(f)Q] = Z(f)$. ∎

We now clarify the relationships between the superHamiltonian property (3.1ii) and the graded Jacobi identity (3.6).

3.71. Theorem. Upon changing (3.1ii) into (3.6) and keeping the rest of the properties (3.1) intact, one obtains an equivelent definition of the superHamiltonian formalism.

Proof. We have seen that (3.6) follows when (3.1ii) is applied to R. Conversely, if (3.6) is satisfied for fixed H and F and arbitrary $R \in C'$, it means that $Z(C') \sim 0$, where $Z = X_{\{H,F\}} - [X_H, X_F]$. By Lemma 3.70 (with $\Phi = id$) we conclude that $Z = 0$. ∎

We now turn to the last topic of this section: affine superHamiltonian operators and associated Lie superalgebras.

3.72. Definition. A stable Lie superalgebra is a free $\mathbf{Z}_2$-graded K-module $\mathcal{G} = K^N = K^{N_0} \oplus K^{N_1}, N = N_0 + N_1$, together with an even multiplication $[\ ,\ ]$

and the grading $p : \mathcal{G} \to \mathbf{Z}_2$ defined by $p(X) = p(X_i) + p(i)$, where $p(i) = \bar{0}$ for $i \leq N_0$ and $p(i) = \bar{1}$ for $i > N_0$, satisfying the following properties:

(3.72i) $[X, Y] = -(-1)^{p(X)p(Y)}[Y, X], \quad \forall X, Y \in \mathcal{G};$

(3.72ii) $[[X, Y], Z] = [X, [Y, Z]] - (-1)^{p(X)p(Y)}[Y, [X, Z]], \quad \forall X, Y, Z \in \mathcal{G};$

(3.72iii) Let $K_c = \{\rho \in K | \widehat{g}(\rho) = \rho, \forall g \in G; \partial_s(\rho) = 0, s = 1, ..., m\}$ be the subring of constants in K. Then $[X, Y\rho] = [X, Y]\rho$, $\forall X, Y \in \mathcal{G}$, $\forall \rho \in K_c$.

(3.72i^4) Multiplication in $\mathcal{G}$ is an operator with respect to each argument, of the following form:

(3.73) $[X, Y]_{\mathrm{k}} = \sum (-1)^{p(i)p(X)} c^{\mathrm{k}}_{i,h|\nu\,;j,g|\sigma}\, \widehat{g}\, \partial^{\sigma}(X_j) \cdot \widehat{h}\, \partial^{\nu}(Y_i), c^{\mathrm{k}}_{\dots} \in K,$

$\forall X, Y \in \mathcal{G}.$

In particular, the sum in (3.73) is finite for each k, $1 \leq \mathrm{k} \leq N$, even if N is infinite. (In the case $N = \infty$, elements of K'^N are finite vectors, i.e., vectors with only a finite number of nonzero components.);

(3.72i^5) The properties (3.72$i - i^4$) remain true under arbitrary extension $K' \supset K$ which makes $\mathcal{G}$ into $\mathcal{G}' = K'^N$:

The formula (3.73) and the property (3.72i^3) of the definition may appear strange and, for the reader familiar with the standard complex Lie superalgebras (see [Ka 2]), even bewildering. The ultimate reason for this definition is that that is what comes out of classifying algebras associated with linear super-Hamiltonian operators (as we shall see below), in complete analogy with Lie algebras which turn out to be in one-to-one correspondence with linear Hamiltonian operators (see [G–Do 3; Ku 9, 4]). I will make just a few comments in order to clarify this definition.

3.74. Proposition. Formulae (3.72i) and (3.73) are compatible.

Proof. First notice that

$$(3.75)\quad p(c^{\mathrm{k}}_{i,h|\nu;j,g|\sigma}) = p(i) + p(j) + p(\mathrm{k}),$$

$$(3.76)\quad c^{\mathrm{k}}_{j,g|\sigma;i,h|\nu} = -(-1)^{p(i)p(j)} c^{\mathrm{k}}_{i,h|\nu\,;j,g|\sigma}.$$

Indeed, (3.75) follows by equating the $\mathbf{Z}_2$-gradings of each side of (3.73) for even X and Y. Similarly, for even X, Y we can use (3.72i) and (3.73) to get (omitting for brevity the G– and ∂–indices):

$$\sum c^{\mathrm{k}}_{ij}(X_j)(Y_i) = [X,Y]_{\mathrm{k}} = -[Y,X]_{\mathrm{k}} = -\sum c^{\mathrm{k}}_{ji}(Y_i)(X_j) =$$
$$= -\sum c^{\mathrm{k}}_{ji}(X_j)(Y_i)(-1)^{p(i)p(j)},$$

and since X and Y can be taken from K'^N for arbitrary $K' \supset K$, (3.76) follows by Lemma 3.19.

Now we show that (3.73) and (3.76) imply (3.72i). We have, again omitting unessential indices,

$$[X,Y]_{\mathrm{k}} = \sum(-1)^{p(i)p(X)} c^{\mathrm{k}}_{ij}(X_j)(Y_i) =$$
$$= \sum(-1)^{p(i)p(X)}(-1)^{p(i)p(j)+1}\, c^{\mathrm{k}}_{ji}\,(Y_i)\,(X_j)(-1)^{[p(Y)+p(i)][p(X)+p(j)]} =$$
$$= \sum(-1)^{p(j)p(Y)} c^{\mathrm{k}}_{ji}(Y_i)(X_j)(-1)^{p(X)p(Y)+1} = -(-1)^{p(X)p(Y)}[Y,X]_{\mathrm{k}}.$$ ∎

3.77. Proposition. The Property (3.72i^3) follows from (3.73).

Proof. No $p(Y)$ enters into (3.73). ∎

3.78. Proposition. If formula (3.73) is satisfied for all even $X, Y \in \mathcal{G}$ and the properties (3.72i, i^3) hold stably, then (3.73) is satisfied for all $X, Y \in \mathcal{G}$.

Proof. Take $\rho = \theta$ from $K'_c \supset K_c[\theta]$ to serve as an odd constant. If Y is odd then $Y\rho$ is even and using (3.72i^3) we find that, for given X, if (3.73) is satisfied for all even Y then it is satisfied for all Y. In particular, it is satisfied for X even and Y odd. Therefore, it remains to consider only the case when X is odd and Y is even. Using (3.76) (which, as we have seen, follows from (3.72i) and {(3.73) for even X and Y}) and (3.72i), we obtain

$$[X,Y]_{\mathrm{k}} = -[Y,X]_{\mathrm{k}} = -\sum c_{ij}^{\mathrm{k}}(Y_j)(X_i) = \sum(-1)^{p(i)p(j)}c_{ji}^{\mathrm{k}}(Y_j)(X_i) =$$
$$= \sum(-1)^{p(i)p(j)}c_{ji}^{\mathrm{k}}(X_i)(Y_j)(-1)^{[1+p(i)]p(j)} =$$
$$= \sum(-1)^{p(j)}c_{ji}^{\mathrm{k}}(X_i)(Y_i) = \sum(-1)^{p(j)p(X)}c_{ji}^{\mathrm{k}}(X_i)(Y_j),$$

which is (3.73) for X odd and Y even. ∎

Thus, the multiplication in a stable Lie superalgebra can be reconstructed from the multiplication in this superalgebra of even elements only . (In 'coordinates', this is evident from (3.73): the structure constants $c^{\mathrm{k}}_{\dots}$ are defined by the products of even elements only.) In a sense, then, the study of stable Lie superalgebras is equivalent to the study of Lie algebras over commutative superalgebras instead of over commutative algebras. (A good exercise is to see how it works for the classical case $K = \mathbf{C}, N < \infty$, see [Ka 2].)

Our plan now is this: to each stable Lie superalgebra we associate a linear (in q's) superHamiltonian matrix, and vice versa; and then we show that affine Hamiltonian matrices are in one-to-one correspondence with generalized 2-cocycles on stable Lie superalgebras.

3.79. <u>Definition</u>. An operator $A : C^a \to C^b$ is linear (in q's) if each of its matrix elements is linear which means having the form $\sum \varphi \hat{g} \partial^\nu$ with $\varphi = \sum c_i^{h|\sigma} q_i^{(h|\sigma)}, c_i^{h|\sigma} \in K$.

Thus, being linear is a stable property.

3.80. <u>Definition.</u> For each stable Lie superalgebra $\mathcal{G}$, let $B^\alpha = B^\alpha(\mathcal{G}), \alpha \in \in$

$\mathbf{Z}_2$, be linear operators defined by the formula

(3.81) $[B^{p(X)}(X)]^t Y \sim \bar{q}^t[X,Y] := \sum q_{\mathrm{k}}[X,Y]_{\mathrm{k}}, \quad \forall\, X,Y \in \mathcal{G}'.$

3.82. Theorem. The matrix $B = B^0$ is correctly defined and is superHamiltonian.

Proof. From (3.73), we have,

$$\bar{q}^{\,t}[X,Y] = \sum q_{\mathrm{k}}(-1)^{p(i)p(X)} c^{\mathrm{k}}_{i,h|\nu;j,g|\sigma}\, \widehat{g}\,\partial^\sigma(X_j)\cdot \widehat{h}\,\partial^\nu(Y_i) \sim$$

$$\sim \sum \left[\widehat{h}^{-1}(-\partial)^\nu(-1)^{p(i)p(X)} q_{\mathrm{k}} c^{\mathrm{k}}_{i,h|\nu;j,g|\sigma}\widehat{g}\,\partial^\sigma(X_j)\right]\cdot Y_i\,.$$

Hence, by Lemma 2.8, $B^{p(X)}(X)$ is uniquely defined which implies, by Lemma 3.19, that $B^{p(X)}$ is uniquely defined as well:

(3.83) $B^\alpha_{ij} = (-1)^{p(i)\alpha}\sum \widehat{h}^{-1}(-\partial)^\nu q_{\mathrm{k}} c^{\mathrm{k}}_{i,h|\nu;j,g|\sigma}\,\widehat{g}\,\partial^\sigma, \quad \alpha\in\mathbf{Z}_2\,.$

In particular, $B^1_{ij} = (-1)^{p(i)}B_{ij}$. We next show that $B^{s\dagger} = -B$. For arbitrary even $X,Y \in K'^N$, we have

$[B^{s\dagger}(Y)]^t X$ [by (2.45)] $\sim [B(X)]^t Y$ [by (3.81)] $\sim \bar{q}^{\,t}[X,Y]$ [by (3.72*i*)] $=$
$= -\bar{q}^{\,t}[Y,X] \sim -[B(Y)]^t X,$

and by Lemma 2.8 we conclude that $B^{s\dagger}(Y) = -B(Y)$, which then forces $B^{s\dagger} = -B$ by Lemma 3.19.

We now check (3.1*ii*). By Theorem 3.52 it is enough to check (3.1*ii*) stably for even linear functions only. We take $H = \bar{q}^{\,t}X, F = \bar{q}^{\,t}Y, R = \bar{q}^{\,t}Z$, with

$p(X) = p(Y) = 0$, and apply both sides of $(3.1ii)$ to R. Let us see that we obtain an identity modulo $Im\mathcal{D}$. First, we notice that

$$(3.84)\quad \{\bar{q}^{\,t}Y, \bar{q}^{\,t}R\} \sim \bar{q}^{\,t}[Y, R], \ \forall Y, R \in K'^N.$$

Indeed, $\frac{\delta}{\delta\bar{q}}(\bar{q}^{\,t}Y) = Y, \frac{\delta}{\delta\bar{q}}(\bar{q}^{\,t}R) = R$. Hence,

$$\{\bar{q}^{\,t}Y, \bar{q}^{\,t}R\} = X_{\bar{q}^{\,t}Y}(\bar{q}^{\,t}R) \sim (\overline{X}_{\bar{q}^{\,t}Y})^t \frac{\delta}{\delta\bar{q}}(\bar{q}^{\,t}R) \ \text{[by (3.2)]}$$

$$= [B^{p(Y)}(Y)]^t R \ \text{[by (3.81)]} \ \sim \bar{q}^{\,t}[Y, R].$$

In particular, from (3.84) we deduce

$$(3.85)\quad \frac{\delta}{\delta\bar{q}}\ \{\bar{q}^{\,t}Y, \bar{q}^{\,t}R\} = [Y, R].$$

Now, for the left-hand-side of $(3.1ii)$ applied to R we obtain

$$(3.86)\quad X_{\{H,F\}}(R) = \{\{H, F\}, R\} \sim \{\bar{q}^{\,t}[X, Y], \bar{q}^{\,t}R\} \sim q^t[[X, Y], R],$$

and for the right-hand-side of $(3.1ii)$ applied to R we get

$$[X_H, X_F](R) = X_H(\{F, R\}) - (-1)^{p(F)p(H)} X_F(\{H, R\}) =$$

$$(3.87)$$

$$= \{H, \{F, R\}\} - (-1)^{p(H)p(F)}\{F, \{H, R\}\} \ \text{[by (3.85)]} \ \sim$$

$$\sim \bar{q}^{\,t}\{[X, [Y, R]] - (-1)^{p(X)p(Y)}[Y, [X, R]]\}.$$

Using $(3.72ii)$ we see that $(3.86) \sim (3.87)$. Thus, $(X_{\{H,F\}} - [X_H, X_F])(R) \sim 0$, and by Lemma 3.70 it follows that $X_{\{H,F\}} - [X_H, X_F] = 0$. i.e., $(3.1ii)$

is satisfied. ∎

The converse to Theorem 3.82 is also true.

3.88. Theorem. Suppose a linear matrix B defines a superHamiltonian structure in C. Define a multiplication $[\ ,\ ]$ in K'^N by the formula (3.81). Then this multiplication is correctly defined and it makes $\mathcal{G} = K'^N$ into a stable Lie superalgebra.

Proof. From the Definition 3.79 it follows that there exists a multiplication $[\ ,\ ]$ in K'^N figuring in (3.81), and from Theorem 2.13(a) applied to $\omega = d(\bar{q}^t[X,Y]) \in \Omega_0^1$ it follows that this multiplication is unique. Taking the $\mathbf{Z}_2$-grading of each part of (3.81) we find that this multiplication is even. Now we check that the properties (3.72) are satisfied. First, $\forall X, Y \in K'^N$,

$$\bar{q}^t[X,Y] \sim [B^{p(X)}(X)]^t Y = \sum(-1)^{p(i)p(X)} B_{ij}(X_j) \cdot Y_i \sim$$

$$\sim \sum(-1)^{p(i)p(X)} B_{ij}^{\dagger}(Y_i) \cdot X_j(-1)^{[p(X)+p(j)][p(Y)+p(i)]} \qquad [\text{by } (3.9)] =$$

$$= \sum(-1)^{[p(X)+p(j)]p(Y)+1} B_{ji}(Y_i) \cdot X_j = -\sum(-1)^{p(X)p(Y)}[B^{p(Y)}(Y)]^t X \sim$$

$$\sim -\bar{q}^t[Y,X](-1)^{p(X)p(Y)},$$

hence (3.72i) follows. Now, since B is linear, we can write

$$(3.89)\quad B_{ij} = \sum q_{\mathrm{k}}^{(h|\nu)} f_{\mathrm{k}ij}^{h|\nu;g|\sigma} \hat{g}\,\partial^\sigma, \quad \textit{finite} \text{ sum for each } (i,j).$$

Therefore,

$$[B^{p(X)}(X)]^t\, Y = \sum(-1)^{p(i)p(X)} B_{ij}(X_j) Y_i \sim$$

$$\sim \sum q_{\mathrm{k}}(-1)^{p(i)p(X)} \hat{h}^{-1}(-\partial)^\nu \left[f_{\mathrm{k}ij}^{h|\nu;g|\sigma} \hat{g}\,\partial^\sigma(X_j) \cdot Y_i\right] \sim \sum q_{\mathrm{k}}[X,Y]_{\mathrm{k}},$$

and (3.73) follows. (Notice, that if X and Y are finite vectors then so is $[X,Y]$ since all the sums in (3.89) are finite.) Hence, (3.72i^3) follows by Proposition 3.77 (or by the law of evidence). The stability property (3.72i^5) is obviously

satisfied, so it remains to check the graded Jacobi identity (3.72ii). Define $H = \overline{q}^t X$, $F = \overline{q}^t Y$, $R = \overline{q}^t Z$. Since $\dfrac{\delta F}{\delta \overline{q}} = Y$, then $\overline{X}_F = B^{p(F)}\left(\dfrac{\delta F}{\delta \overline{q}}\right) =$
$= B^{p(Y)}(Y)$.

Therefore,

$$\{F, R\} = X_F(R) \sim < X_F, \delta(R) >= \overline{X}_F^t Z = [B^{p(Y)}(Y)]^t Z \sim q^t[Y, Z],$$

which is (3.84). Now, substituting our H, F, and R into (3.6) and using (3.84) we obtain

$$(3.90)\quad \overline{q}^t\,[[X, Y], Z] \sim \overline{q}^t\left\{[X, [Y, Z]] - (-1)^{p(X)p(Y)}[Y, [X, Z]]\right\}.$$

Therefore, (3.72ii) follows by Theorem 2.13(a) applied to the differential forms resulting in acting by the differential d on each side of the relation (3.90). ∎

3.91. Remark. Since the map between linear superHamiltonian structures and stable Lie superalgebras is given, in each direction, by the same formula (3.81), we have, in fact, a one-to-one correspondence.

Now we turn to affine superHamiltonian operators.

3.92. Definition. A bilinear form on K^N is a stable map $K'^N \times K'^N \to K'$ which is an operator with coefficients in K with respect to each argument, and is a right K'_c-homomorphism with respect to the second argument, $\forall K' \supset K$.

3.93. Definition. Two bilinear forms ω_1 and ω_2 are equivalent if $\omega_1(X, Y) \sim \omega_2(X, Y)$, $\forall X, Y \in K'^N$.

3.94. Corollary. Every bilinear form is equivalent to a form of the type

$$(3.95)\quad \omega(X, Y) \sim \sum b^{p(X)}_{i;j,g|\sigma}\, \widehat{g}\, \partial^\sigma(X_j) \cdot Y_i, \quad b^{p(X)}_{\dots} \in K,$$

and such representation is unique.

3.96. Definition. A bilinear form ω is called superskewsymmetric if it satisfies the relation

$$(3.97)\quad \omega(X,Y) \sim -(-1)^{p(X)p(Y)}\omega(Y,X), \quad \forall X, Y \in K'^N.$$

3.98. Definition. An operator b with matrix elements $b_{ij} = \sum b^0_{i;j,g|\sigma}\widehat{g}\partial^\sigma$ defined by (3.95) is called associated with (or corresponding to) the bilinear form ω.

3.99. Lemma. If ω is a bilinear superskewsymmetric form and b is the associated operator then $p(b) = p(\omega)$,

$$(3.100)\quad \omega(X,Y) \sim \sum (-1)^{p(i)p(X)} b_{ij}(X_j) \cdot Y_i\,,$$

and b is superskewsymmetric.

Proof. The superskewsymmetric property of ω, written in longhand as

$$(-1)^{p(X)p(Y)+1} \sum b^{p(Y)}_{ji}(Y_i) \cdot X_j \sim -(-1)^{p(X)p(Y)}\omega(Y,X) \sim \omega(X,Y) \sim$$

(3.101)

$$\sim \sum b^{p(X)}_{ij}(X_j) \cdot Y_i \sim \sum \left(b^{p(X)}_{ij}\right)^\dagger (Y_i) \cdot X_j (-1)^{[p(X)+p(j)][p(Y)+p(i)]}\,,$$

is equivalent to

$$(3.102)\quad b^{p(Y)}_{ji} = \left(b^{p(X)}_{ij}\right)^\dagger (-1)^{1+p(i)p(j)+p(Y)p(j)+p(X)p(i)}\,.$$

For $p(X) = p(Y) = 0$, (3.102) yields

$$(3.103)\quad (b_{ij})^\dagger = -b_{ji}(-1)^{p(i)p(j)}\,.$$

For $p(X) = 0$, (3.102) together with (3.103) imply

$$(3.104)\quad b^{p(Y)}_{ji} = (-1)^{p(j)p(Y)} b_{ji}.$$

Taking the adjoint of (3.104) and using (3.103) we obtain

(3.105) $\left(b_{ij}^{p(X)}\right)^{\dagger} = (-1)^{p(i)p(X)+p(i)p(j)+1} b_{ji}.$

Substituting (3.104) in (3.102) results in an identity. Thus, the equality (3.102) is equivalent to the pair of equalities (3.104) and (3.103), the first of which proves (3.100) and the second one proves that b is superskewsymmetric. Finally, taking the $\mathbf{Z}_2$-gradings of both parts of (3.100) we obtain

$$p(\omega) + p(X) + P(Y) = p(b_{ij}) + p(i) + p(j) + p(X) + p(Y),$$

so that

(3.106) $p(b) = p(b_{ij}) + p(i) + p(j) = p(\omega).$ ∎

3.107. Lemma. Let b be superskewsymmetric. Then the bilinear form ω defined by (3.100) is superskewsymmetric and $p(\omega) = p(b)$.

Proof. From (3.100) follows (3.106) which shows that $p(\omega) = p(b)$. To prove the rest of the Lemma, start with (3.103); use (3.104) and (3.103) to deduce (3.102); and use (3.102) to establish (3.97) by walking over the chain (3.101) backwards. ∎

3.108. Definition. A (generalized) two-cocycle on a stable Lie superalgebra $\mathcal{G} = K'^N$ is an even superskewsymmetric form ω satisfying

(3.109) $\omega([X,Y],Z) \sim \omega(X,[Y,Z]) - (-1)^{p(X)p(Y)}\omega(Y,[X,Z]),$

$\forall X, Y, Z \in \mathcal{G}.$

3.110. Theorem. Let ω be a two-cocycle on a stable Lie superalgebra $\mathcal{G}$, and let $b = b_\omega$ be the operator corresponding to ω. Then the matrix

(3.111) $\widetilde{B} = B(\mathcal{G}) + b_\omega$

is superHamiltonian.

Proof. By Theorem 3.82, $B(\mathcal{G})$ is superHamiltonian, and, in particular, it is even superskewsymmetric. By Lemma 3.99, b_ω is even superskewsymmetric too, therefore $\widetilde{B}$ is even superskewsymmetric as well. To check (3.1*ii*) we follow the Proof of Theorem 3.82: By Theorem 3.52 we can take H and F being linear, and then we make sure that after applying each side of (3.1*ii*) to $R = \bar{q}^t Z$ we obtain an equality modulo $Im\mathcal{D}$. So, let $H = \bar{q}^t X$, $F = \bar{q}^t Y$, with $X, Y, Z \in K'^N$. Then $\left(\frac{\delta F}{\delta \bar{q}}\right) = Y$, and

$$\text{(3.112)} \qquad \{\bar{q}^t Y, \bar{q}^t R\} = X_{\bar{q}^t Y}(\bar{q}^t R) \sim < \overline{X}_{\bar{q}^t Y}, \delta(\bar{q}^t R) >=$$

$$= [\widetilde{B}^{p(Y)}(Y)]^t R = [B^{p(Y)}(Y)]^t R + [b^{p(Y)}(Y)]^t R \sim \bar{q}^t[Y, R] + \omega(Y, R).$$

In particular,

$$\text{(3.113)} \quad \frac{\delta}{\delta \bar{q}}\{\bar{q}^t Y, \bar{q}^t R\} = [Y, R].$$

Therefore, for the left-hand-side of (3.1*ii*) applied to R we get

$$X_{\{H,F\}}(R) = \{\{H, F\}, R\} \sim \{\bar{q}^t[X, Y] + \omega(X, Y), \bar{q}^t Z\} =$$

$$\text{(3.114)} \qquad = \{\bar{q}^t[X, Y], \bar{q}^t Z\} \text{ [by (3.112)]} \sim$$

$$\sim \bar{q}^t[[X, Y], Z] + \omega([X, Y], Z),$$

while for the right-hand-side of (3.1*ii*) applied to R we obtain

$$[X_H, X_F](R) = X_H(\{F, R\}) - (-1)^{p(H)p(F)} X_F(\{H, R\}) =$$

$$\text{(3.115)} \quad = \{H, \{F, R\}\} - (-1)^{p(H)p(F)}\{F, \{H, R\}\} \text{[by (3.112)]} \sim$$

$$\sim \{\bar{q}^t X, \bar{q}^t[Y, Z] + \omega(Y, Z)\} - (-1)^{p(X)p(Y)}\{\bar{q}^t Y, q^t[X, Z] + \omega(X, Z)\} \sim$$

$$\sim \bar{q}^t[X, [Y, Z]] + \omega(X, [Y, Z]) - (-1)^{p(X)p(Y)} \left(\bar{q}^t[Y, [X, Z]] + \omega(Y, [X, Z])\right).$$

Since $\mathcal{G}$ is a stable Lie superalgebra and ω is a 2-cocycle on $\mathcal{G}$, (3.114) is equivalent to (3.115) by (3.72*ii*) and (3.109). ∎

Finally, we prove the converse to Theorem 3.110.

3.116. Theorem. Suppose $\widetilde{B} = B + b$ is a superHamiltonian matrix, where B is an even linear superskewsymmetric matrix and b is an even superskewsymetric matrix with coefficients in K. Define a multiplication [,] in K'^N by the formula (3.81). Then this multiplication is correctly defined, makes $\mathcal{G} = K'^N$ into a stable Lie superalgebra, and the bilinear form ω on $\mathcal{G}$ defined by the formula (3.100) is a 2-cocycle on $\mathcal{G}$.

Proof. We first show that B itself is superHamiltonian, given that $B + b$ is. Keeping the notation of the Proof of Theorem 3.110, we use (3.112–3.115) to find that (3.1*ii*) applied to R becomes

$$(3.117) \quad \overline{q}^{\,t} a + \epsilon \sim 0 ,$$

where

$$(3.118) \quad a = [[X,Y],Z] - ([X,[Y,Z]] - (-1)^{p(X)p(Y)}[Y,[X,Z]]),$$

$$(3.119) \quad \epsilon = \omega([X,Y],Z) - (\omega(X,[Y,Z]) - (-1)^{p(X)p(Y)}\omega(Y,[X,Z])) .$$

From (3.117) we obtain $0 = \dfrac{\delta}{\delta \overline{q}}(\overline{q}^{\,t} a + \epsilon) = a$, so that $a = 0$, which means that B is superHamiltonian. By Theorem 3.88, $\mathcal{G}$ is a Lie superalgebra. Now, since $a = 0$ in (3.117) we find that $\epsilon \sim 0$, and (3.119) then implies that ω is a two-cocycle on $\mathcal{G}$ ∎

3.120. Remark. Since the map between affine superHamiltonian operators and two-cocycles on stable Lie superalgebras is given, in each direction, by the same formulae (3.81) and (3.100), this map is, in fact, a one-to-one correspondence.

§4. Residue Calculus in Modules of Differential Forms over Superalgebras of Pseudo-Differential Operators

The main notions of this section are: the supertrace; pseudo-differential operators; the differential of a matrix pseudo-differential operator; the Residue. The main results are: supersymmetry and invariance of the form 'supertrace of the Residue of the product' on the Lie superalgebra of matrix pseudo-differential operators; formula for the supertrace of the Residue of the differential of a power of a matrix pseudo-differential operator.

We start by recalling the properties of the supertrace (see [Ka 2]). Let T be a commutative superalgebra, E be a $\mathbf{Z}_2$-graded associative ring and left T-module, and E' be a $\mathbf{Z}_2$-graded left E-module. We fix two nonnegative integers ℓ_0 and ℓ_1 at least one of which is positive. Set $\ell = \ell_0 + \ell_1$. Let $Mat_\ell(E')$ be the set of all ℓ x ℓ matices with entries in E'. We introduce the following $\mathbf{Z}_2$-grading on $Mat_\ell(E')$:

(4.1) $p(a) = p(a_{\alpha\beta}) + p(\alpha) + p(\beta), \quad a \in Mat_\ell(E'),$

where

(4.2) $p(\alpha) = 0, \alpha \leq \ell_0; \; p(\alpha) = 1, \; \alpha > \ell_0.$

4.3. Proposition. The multiplication map: $Mat_\ell(E)$ x $Mat_\ell(E') \rightarrow$
$\rightarrow Mat_\ell(E')$ is even.

Proof. We have to show that

$$p(ab) = p(a) + p(b), \quad \forall a \in Mat_\ell(E), \; \forall b \in Mat_\ell(E').$$

By (4.1),

$p(a) + p(b) = p(a_{\alpha\gamma}) + p(\alpha) + p(\gamma) + p(b_{\gamma\beta}) + p(\gamma) + p(\beta) =$

$= p(a_{\alpha\gamma}b_{\gamma\beta}) + p(\alpha) + p(\beta) = p[(ab)_{\alpha\beta}] + p(\alpha) + p(\beta) = p(ab).$ ∎

4.4. Definition. The supertrace of a matrix is defined as

(4.5) $str(a) = \sum_{\alpha=1}^{\ell} (-1)^{p(\alpha)[1+p(a)]} a_{\alpha\alpha}, \quad a \in Mat_\ell(E').$

4.6. Lemma. If T is a commutative superalgebra then

$$(4.7)\quad str\,(ab) = (-1)^{p(a)p(b)}\, str(ba), \quad \forall a, b \in Mat_\ell(\mathrm{T}).$$

Proof. We have,

$$(4.8)\quad str(ab) = \sum (-1)^{p(\gamma)[1+p(ab)]}(ab)_{\gamma\gamma} = \sum (-1)^{p(\gamma)[1+p(ab)]} a_{\gamma\alpha} b_{\alpha\gamma} =$$

$$= \sum (-1)^{p(\gamma)[1+p(a)+p(b)]} b_{\alpha\gamma} a_{\gamma\alpha}\, (-1)^{[p(a)+p(\gamma)+p(\alpha)][p(b)+p(\alpha)+p(\gamma)]},$$

$$(4.9)\quad (-1)^{p(a)p(b)}\, str(ba) = \sum (-1)^{p(a)p(b)}\, (-1)^{p(\alpha)[1+p(b)+p(a)]}\, b_{\alpha\gamma}\, a_{\gamma\alpha},$$

and since

$$p(\gamma)[1+p(a)+p(b)] + [p(a)+p(\gamma)+p(\alpha)][p(b)+p(\alpha)+p(\gamma)] =$$

$$= p(a)p(b) + p(\alpha)[1+p(b)+p(a)],$$

we conclude that (4.8) and (4.9) are equal. ∎

4.10. Remark. $Mat_\ell(E)$ generates a Lie superabgebra $Mat_\ell(E)^{Lie}$ via the commutator

$$(4.11)\quad [a, b] = ab - (-1)^{p(a)p(b)}\, ba.$$

In the language of Lie superalgebras, an equivalent form of Lemma 4.6 is this:

4.12. Lemma. If T is a commutative superalgebra than

$$(4.13)\quad str([a, b]) = 0, \qquad \forall a, b \in Mat_\ell(\mathrm{T})^{Lie}.$$

Recall that if E' is a left E-module it can be also considered as a E-bimodule, via the rule

$$(4.14)\quad ba = (-1)^{p(a)p(b)}\, ab, \qquad a \in E, b \in E'.$$

In particular, the E-bimodule structure of E itself is compatible with the ring structure of E if and only if E is a commutative superalgebra.

4.15. Lemma. If E' is an E-bimodule then

(4.16) $str(ab) = (-1)^{p(a)p(b)}\, str(ba),\ \forall a \in Mat_\ell(E), b \in Mat_\ell(E')$.

Proof is exactly the same as that of Lemma 4.6. ∎

We now turn to the construction of rings and modules of pseudo-differential operators.

Suppose k and K are as in §§2,3. We suppose, for the remainder of this Chapter, *that the number of derivations* m *is* ≤ 1, that is, that there is either only one derivation $\partial = \partial_1$ present or none at all. We assume that ∂ *is* present; in the case it is not, all the differential indices in the formulae below should be dropped off with the remaining (group) indices left intact.

Suppose that G and ∂ also act on E and E', that T $= K$, that E is a K-bimodule, and that the actions of G and ∂ on K, E, and E' are compatible with the K-binmodule structure of E and E-bimodule structure of E'.

4.17. Definition. A pseudo-differential operator with coefficients in E' is an expression of the form

(4.18) $A = \sum a_{g|\nu}\, \hat{g}\, \xi^\nu, \quad a_{g|\nu} \in E', g \in G,\ \nu \in \mathbf{Z}$,

where the range of summation in (4.18) satisfies the following conditions:

(4.19.∂) $\nu \leq \bar{\nu}(A) < \infty$,

(4.19.G) For each $\nu, a_{g|\nu}$ is nonzero for only a finite number of $g \in G$.

(The property (4.19.G) can be sometimes weakened; e.g., if G is a $\mathbf{Z}$-graded group then (4.19.G) can be exchanged for:

(4.19.G$'$) for each ν, there exists only a finite number of elements $g \in G$ of positive $\mathbf{Z}$-grading for which $a_{g|\nu}$ is nonzero.)

We define the $\mathbf{Z}_2$-grading of a pseudo-differential operator by the formula

(4.20) $p(a_{g|\nu}\, \hat{g}\, \xi^\nu) = p(a_{g|\nu})\,,\ a_{g|\nu} \in E'$.

The set of all pseudo-differential operators with coefficients in E′ is denoted $\mathcal{O}_{E'}$. It is a $\mathbf{Z}_2$-graded, natural left E-module. We now make $\mathcal{O}_{E'}$ into a left $\mathcal{O}_E$-module.

4.21. Definition. Set

$$(4.22)\quad \widehat{g}\,\xi^n = \xi^n\,\widehat{g}, \qquad n \in \mathbf{Z}, g \in G,$$

$$(4.23)\quad (a\xi^r\widehat{g})(b\widehat{h}\xi^n) = (a\xi^r)\widehat{g}(b)\widehat{gh}\,\xi^n, \quad a \in E,\ \ b \in E',\ \ h \in G,\ \ r \in \mathbf{Z},$$

$$(4.24)\quad (a\xi^n)(b\xi^r) = a\sum_{\mathrm{k}\geq 0}\binom{n}{\mathrm{k}} b^{(\mathrm{k})}\xi^{n-\mathrm{k}+r}, \qquad b^{(\mathrm{k})} := \partial^{\mathrm{k}}(b).$$

4.25. Lemma. With the multiplication rules (4.22)–(4.24), $\mathcal{O}_{E'}$ is a left $\mathcal{O}_E$-module. In particular, $\mathcal{O}_E$ itself is an associative ring.

Proof. We have to show that

$$(4.26)\quad [(a'\widehat{g}\xi^n)(a\widehat{h}\xi^r)](b\widehat{g}'\xi^{\mathrm{k}}) = (a'\widehat{g}\xi^n)[(a\widehat{h}\xi^r)(b\widehat{g}'\xi^{\mathrm{k}})].$$
$$\forall a, a' \in E, \quad \forall b \in E', \quad \forall g, h, g' \in G, \quad \forall n, r, \mathrm{k} \in \mathbf{Z}.$$

We can get rid of $\widehat{g}, \widehat{h}, \widehat{g}', a'$, and ξ^{k}, since they do not contribute any problem to checking (4.26), as is clear from (4.22)–(4.24). Thus, we have to check only that

$$(4.27)\quad [\xi^n(a\xi^r)]b = \xi^n[(a\xi^r)b].$$

For the left-hand-side of (4.27) we obtain

$$(4.28)\quad \left[\sum_{\gamma\geq 0}\binom{n}{\gamma}a^{(\gamma)}\xi^{n+r-\gamma}\right]b = \sum_{\gamma,\mu\geq 0} a^{(\gamma)}b^{(\mu)}\binom{n}{\gamma}\binom{n+r-\gamma}{\mu}\xi^{n+r-\gamma-\mu},$$

while for the right-hand-side of (4.27) we get

$$(4.29)\quad \xi^n\left[a\sum_{\alpha\geq 0}\binom{r}{\alpha}b^{(\alpha)}\xi^{r-\alpha}\right] = \sum_{\alpha,\beta\geq 0}(ab^{(\alpha)})^{(\beta)}\binom{n}{\beta}\binom{r}{\alpha}\xi^{n+r-\alpha-\beta} =$$

$$= \sum_{\alpha,\beta,\gamma\geq 0} a^{(\gamma)}b^{(\beta+\alpha-\gamma)}\binom{\beta}{\gamma}\binom{n}{\beta}\binom{r}{\alpha}\xi^{n+r-\alpha-\beta}.$$

Hence, (4.27) is equivalent to the equality

$$(4.30)\quad \binom{n}{\gamma}\binom{n+r-\gamma}{\mu} = \sum_{\substack{\alpha+\beta=\gamma+\mu\\ \alpha,\beta\geq 0}} \binom{\beta}{\gamma}\binom{n}{\beta}\binom{r}{\alpha}. \qquad n,r\in\mathbf{Z},\gamma,\mu\in\mathbf{Z}_+.$$

To prove (4.30), we start with the obvious identity

$$(4.31)\quad \gamma!\binom{n}{\gamma}(1+x)^{n-\gamma}(1+y)^r\Big|_{y=x} = \left[\frac{1}{\gamma!}\left(\frac{\partial}{\partial x}\right)^{\gamma}(1+x)^n(1+y)^r\right]\Big|_{y=x}.$$

Picking out the x^μ-coefficients from both parts of (4.31), we obtain

$$\binom{n}{\gamma}\binom{n+r-\gamma}{\mu} = \frac{1}{\gamma!}\left\{x^\mu - coef.of\left\{\left[\frac{1}{\gamma!}\left(\frac{\partial}{\partial x}\right)^{\gamma}\sum\binom{n}{\beta}x^\beta\binom{r}{\alpha}y^\alpha\right]\Big|_{y=x}\right\}\right\} =$$

$$= x^\mu - coef.\ of\left\{\left[\binom{\beta}{\gamma}\sum\binom{n}{\beta}\binom{r}{\alpha}x^{\beta-\gamma}x^\alpha\right]\right\} = \sum_{\alpha+\beta=\gamma+\mu}\binom{\beta}{\gamma}\binom{n}{\beta}\binom{r}{\alpha}. \blacksquare$$

4.32. Remarks. 1) The identity (4.27) is well known (see, e.g. [Man]). The Proof above is, I believe, the shortest possible. 2) For $n=1$, (4.24) yields

$$(4.33)\quad \xi b = b\xi + \partial(b).$$

Thus, ξ can be thought of as ∂ itself, but considered as an operator. The name "pseudo-differential" reflects the admission of negative powers of ξ. 3) The same formulae (4.22)-(4.24) make $\mathcal{O}_{E'}$ into a right $\mathcal{O}_E$-module and into a $\mathcal{O}_E$-bimodule. The difference with linear algebra is that the $\mathcal{O}_E$-bimodule structure on $\mathcal{O}_{E'}$ is *not* provided by the formula (4.14). Equivalently, if K is a commutative superalgebra then $\mathcal{O}_K$ is an associative superalgebra but not a commutative superalgebra.

We shall need the following Lemma.

4.34. Lemma.

$$(4.35)\quad b\xi^n = \sum_{k\geq 0}\xi^{n-k}(-1)^k\binom{n}{k}b^{(k)}.$$

Proof. We have,

$$\sum\xi^{n-k}(-1)^k\binom{n}{k}b^{(k)} = \sum\binom{n-k}{\alpha}(-1)^k\binom{n}{k}b^{(k+\alpha)}\xi^{n-k-\alpha}.$$

and, thus, (4.35) follows from the following equality

$$\sum_{k+\alpha=\mu} \binom{n-k}{\alpha} (-1)^k \binom{n}{k} = \delta^{\mu}_{0},$$

or

$$\sum_{k=0}^{\mu} \binom{n-k}{\mu-k} (-1)^k \binom{n}{k} = \delta^{\mu}_{0},$$

or, upon denoting $\eta = n - \mu$,

$$(4.36) \quad \sum_{k\geq 0} \binom{n-k}{\eta} (-1)^k \binom{n}{k} = \delta^{n}_{\eta},$$

where δ is the Kronecker delta. This identity results, in turn, by picking out the x^η-coefficients from the following obvious identity:

$$x^n = [(x+1)-1]^n = \sum_{k\geq 0} \binom{n}{k} (x+1)^{n-k} (-1)^k =$$

$$= \sum_{k,\eta\geq 0} \binom{n}{k} \binom{n-k}{\eta} x^\eta (-1)^k. \qquad \blacksquare$$

4.37. Definition. The Residue of a matrix pseudo-differential operator is given by the formula

$$(4.38) \quad Res\left(\sum a_{g|\nu} \widehat{g} \xi^\nu\right) = a_{e|-1}, \qquad a_{g|\nu} \in Mat_\ell(E').$$

4.39. Theorem. Let $U \in Mat_\ell(\mathcal{O}_E)$, $V \in Mat_\ell(\mathcal{O}_{E'})$. If E' is an E-bimodule then

$$(4.40) \quad str\, Res\,(UV) \sim str\, Res\,(VU)\,(-1)^{p(U)p(V)}.$$

Proof. It is enough to consider the case when U is of the from $a\widehat{g}\xi^n$ and V, thanks to Lemma 4.34, is of the form $\xi^r \widehat{h} b$.

Then

$$Res\,(UV) = Res\,(a\widehat{g}\xi^n \xi^r \widehat{h} b) = ab\delta^{n+r}_{-1}\,\delta^{gh}_{e},$$

$$Res\,(VU) = Res\,(\xi^r \widehat{h} ba \widehat{g} \xi^n) = \delta_e^{hg}\, Res \left[\sum_{k \geq 0} \binom{r}{k} \partial^k \widehat{h}(ba) \cdot \xi^{n+r-k} \right] \sim$$

$$\sim \delta_e^{hg}\, Res \left[\widehat{h}(ba) \xi^{n+r} \right] = \delta_e^{hg} \delta_{-1}^{n+r} \widehat{h}(ba) \sim ba \delta_e^{hg} \delta_{-1}^{n+r},$$

and (4.40) follows from Lemma 4.15 and (4.20). ∎

4.41. Corollary. Consider the following expression

$$(4.42) \quad \omega(U,V) = str\, Res\,(UV), \quad U,V \in Mat_\ell(\mathcal{O}_K).$$

Then ω is a bilinear supersymmetric form on $Mat_\ell(\mathcal{O}_K)$ which is invariant on $Mat_\ell(\mathcal{O}_K)^{Lie}$:

$$(4.43) \quad \omega(U,V) \sim \omega(V,U)\,(-1)^{p(U)p(V)},$$

$$(4.44) \quad \omega\,([U,V],W) \sim \omega\,(U,[V,W]) .$$

Proof. That ω is a bilinear form follows from Definition 3.92 and the computation in the Proof of Theorem 4.39. Formula (4.43) is the same as (4.40). As for (4.44), we have

$$\begin{aligned} \omega\,([U,V],W) &= \omega(UV - (-1)^{p(U)p(V)}\, VU,\, W) = \\ &= str\, Res\, \left([UV - (-1)^{p(U)p(V)}\, VU]\, W \right) [\text{by } (4.40)] \sim \\ &\sim str\, Res\, \left(UVW - (-1)^{p(U)p(V)}\, UWV (-1)^{p(V)p(UW)} \right) = \\ &= str\, Res\, (U\,[V,W]) = \omega\,(U,\, [V,\, W]) . \end{aligned}$$

∎

We are now in a good position to consider the case we were aiming at: $E = C = K\left[q_i^{(g|\mu)}\right]$, $E' = \Omega^1(C)$. (We continue to use notation of §§2, 3; remember that $m = 1$ now, so that $\mu \in \mathbf{Z}_+$.)

4.45. Definition. The map

$$d : \mathcal{O}_C \to \mathcal{O}_{\Omega^1(C)}$$

is given by the formula

(4.46) $d(\xi^r \widehat{g} a \widehat{h} \xi^{\mathrm{k}}) = \xi^r \widehat{g} d(a) \widehat{h} \xi^{\mathrm{k}}, \quad a \in C, \quad g, h \in G, \quad r, \mathrm{k} \in \mathbf{Z}.$

4.47. Lemma. The map d in (4.46) is correctly defined.

Proof. We have to check that the relations (4.22) – (4.24) are compatible with the formula (4.46). For the group action, we have
$d(\widehat{g}a) = d[\widehat{g}(a)\widehat{g}]$ [by (4.46)] $= d(\widehat{g}(a))\widehat{g}$ [since $\widehat{g}$ commutes with $d : C \to \Omega^1(C)$] $= \widehat{g}(d(a))\widehat{g} = \widehat{g}d(a)$.

Similarly,

$$d(\xi^r a) = d\left(\sum_{\mathrm{k}} a^{(\mathrm{k})} \binom{r}{\mathrm{k}} \xi^{r-\mathrm{k}}\right) = \sum d\left(a^{(\mathrm{k})}\right) \binom{r}{\mathrm{k}} \xi^{r-\mathrm{k}} \quad \text{[since } \partial$$

commutes with $d : C \to \Omega^1(C)] = \sum \partial^{\mathrm{k}}(d(a)) \binom{r}{\mathrm{k}} \xi^{r-\mathrm{k}} = \xi^r d(a).$

4.48. Lemma.

(4.49) $d(UV) = d(U)V + (-1)^{p(U)} U d(V), \quad \forall U, V \in \mathcal{O}_C.$

Proof. We have, when U and V are monomials:

$$d[a\widehat{g}\xi^r a' \widehat{h} \xi^n] = d[a \sum_{\mathrm{k}} \binom{r}{\mathrm{k}} \widehat{g} \partial^{\mathrm{k}}(a') \widehat{g} \widehat{h} \xi^{r-\mathrm{k}+n}] =$$

$$\sum_{\mathrm{k}} [d(a) \cdot \widehat{g}\partial^{\mathrm{k}}(a') + (-1)^{p(a)} a \widehat{g} \partial^{\mathrm{k}} d(a')] \binom{r}{\mathrm{k}} \widehat{g}\widehat{h} \xi^{r-\mathrm{k}+n} =$$

$$= d(a)\widehat{g}\xi^r a' \widehat{h}\xi^n + (-1)^{p(a)} a \widehat{g} \xi^r d(a') \widehat{h} \xi^n =$$

$$= d(U) \cdot V + (-1)^{p(U)} U d(V).$$

4.50. Definition. The maps $d : Mat_\ell(C) \to Mat_\ell(\Omega^1(C))$ and $d : Mat_\ell(\mathcal{O}_C) \to Mat_\ell(\mathcal{O}_{\Omega^1(C)})$ are defined as

(4.51) $[d(U)]_{\alpha\beta} = (-1)^{p(\alpha)} d(U_{\alpha\beta}), \quad U \in Mat_\ell(C \text{ or } \mathcal{O}_C).$

4.52. Lemma.

$$(4.53)\quad d(UV) = d(U)\cdot V + (-1)^{p(U)}Ud(V), \qquad \forall U,V \in Mat_\ell(C \text{ or } \mathcal{O}_C).$$

Proof. We have

$$[d(UV)]_{\alpha\beta} = (-1)^{p(\alpha)}\, d[(UV)_{\alpha\beta}] = (-1)^{p(\alpha)}\, d(\sum_\gamma U_{\alpha\gamma}V_{\gamma\beta})\,[\text{by } (4.49)] =$$

$$= (-1)^{p(\alpha)} \sum [d(U_{\alpha\gamma})\cdot V_{\gamma\beta} + (-1)^{p(U_{\alpha\gamma})}\, U_{\alpha\gamma}\, d(V_{\gamma\beta})] =$$

$$= \sum [d(U)]_{\alpha\gamma} V_{\gamma\beta} + \sum (-1)^{p(U)+p(\gamma)}\, U_{\alpha\gamma}[d(V)]_{\gamma\beta}(-1)^{p(\gamma)} =$$

$$= [d(U)\cdot V]_{\alpha\beta} + (-1)^{p(U)}[Ud(V)]_{\alpha\beta}. \qquad \blacksquare$$

4.54. Remark. When $l = l_1 = 1$, the *matrix* map d of the definition 4.50 is *minus the standard d*. Such a strange outcome is a typical feature of the super-calculus. (Even a more bizarre relation is the formula (4.59) below.) However, we have the following result relating the matrix and the scalar differentials:

4.55. Theorem. The following formulae hold:

$$(4.56)\quad str\,[d(a)] = d[str\,(a)], \qquad a \in Mat_\ell(C),$$

$$(4.57)\quad str\ Res\,[d(U)] = d\,[str\ Res\,(U)], \qquad U \in Mat_\ell(\mathcal{O}_C).$$

Proof. If $U = \sum a_{g|\nu}\hat{g}\xi^\nu \in Mat_\ell(\mathcal{O}_C)$, then

$$str\ Res\,[d(U)] = str\,[d(a_{e|-1})]\,[\text{by } (4.56)] = d[str\,(a_{e|-1})] =$$

$$= d[str\ Res\,(U)],$$

so that (4.57) follows from (4.56). Now use (4.5) and (4.51) to obtain

$$str\,[d(a)] = \sum_\alpha (-1)^{p(\alpha)[1+p(d(a))]}\,[d(a)]_{\alpha\alpha} =$$

$$= \sum (-1)^{p(\alpha)p(a)}\,(-1)^{p(\alpha)}\,d(a_{\alpha\alpha}) = d(\sum (-1)^{p(\alpha)[1+p(a)]}a_{\alpha\alpha}) = d[str\,(a)]. \qquad \blacksquare$$

We now derive the main result of this section.

4.58. Theorem. For any $U \in Mat_\ell(\mathcal{O}_C)$, and any $n \in \mathbf{Z}_+$,

$$str\ Res\,[d(U^{n+1})] \sim \begin{cases} 0, \text{ when } p(U)=1 \text{ and } n\equiv 1(\text{mod } 2), & (4.59)\\ (n+1)\,str\ Res\,[d(U)\cdot U^n], \text{ otherwise.} & (4.60)\end{cases}$$

Proof. We have, by (4.53) and using $p(U^{\mathrm{k}}) = \mathrm{k}p(U)$,

$$(4.61)\quad d(U^{n+1}) = \sum_{\mathrm{k}=0}^{n}(-1)^{\mathrm{k}p(U)}U^{\mathrm{k}}d(U)\cdot U^{n-\mathrm{k}}.$$

Taking *str Res* of both sides of (4.61) and using (4.40), we obtain

$$str\ Res\,[d(U^{n+1})] = \sum_{\mathrm{k}=0}^{n}(-1)^{kp(U)}\,str\ Res\,[U^{\mathrm{k}}d(U)U^{n-\mathrm{k}}] \sim$$

$$\sim \sum_{\mathrm{k}=0}^{n}(-1)^{\mathrm{k}p(U)}str\ Res\,[d(U)U^{n-\mathrm{k}}U^{\mathrm{k}}](-1)^{\mathrm{k}p(U)[1+p(U)+(n-\mathrm{k})p(U)]} =$$

(4.62)

$$= \sum_{\mathrm{k}=0}^{n} str\ Res\,[d(U)U^n](-1)^{p(U)\mathrm{k}(n-\mathrm{k}+1)} = str\ Res\,[d(U)U^n]\sum_{\mathrm{k}=0}^{n}(-1)^{p(U)\mathrm{k}n}.$$

Let us compute the sum in the right-hand-side of (4.62):

$$\sum_{\mathrm{k}=0}^{n}(-1)^{p(U)\mathrm{k}n} = \begin{cases} n+1, \text{ if } p(U)=0 \text{ or } n\equiv 0(mod\,2)\\ \sum_{\mathrm{k}=0}^{n}(-1)^{\mathrm{k}} = 0, \text{ otherwise,}\end{cases}$$

and this yields (4.60) and (4.59). ∎

The following result is also useful.

4.63. Lemma.

$$(4.64)\quad str\ Res\,[rU^r d(U)^s)] \sim str\ Res\,[sU^s d(U^r)],$$

$U \in Mat_\ell(\mathcal{O}_C)$, $\quad p(U)=0$, $\quad r,s\in \mathbf{Z}_+$.

Proof. If either r or s vanishes then (4.64) becomes $0\sim 0$. Suppose both r and s are positive. Then $d(U^s) = \sum_{\mathrm{k}=0}^{s-1} U^{\mathrm{k}}d(U)U^{s-\mathrm{k}-1}$, so

$$str\ Res\,[rU^r d(U^s)] \sim str\ Res\,[r\sum_{k=0}^{s-1} d(U)U^{s-k-1}U^{r+k}] =$$

$$= rs\, str\ Res[d(U)U^{s+r-1}],$$

and this expression is symmetric in r and s. ∎

4.65. Remark. Allowing $p(U) = 1$ in Lemma 4.63 leads to a multitude of unenlightening possibilities, and this is why it is better to avoid this case.

§5. Classical Superintegrable Systems

In this section we construct classical superintegrable systems and find their basic properties. The main objects are: a pseudo–differential Lax operator L (5.5); the Lax equations (5.10); the centralizer $Z(L)$ of L (Definition 5.7); admissible elements (Definition 5.15). The main results are: the full description of $Z(L)$ (Theorem 5.36); the supercommutativity of Lax derivations (Theorem 5.80); a construction of an infinite common set of conservation laws for Lax derivations (Theorem 5.81).

Let k and K be as in §§2-4, with G absent and the number of derivations m being one; we again denote ∂_1 by ∂. As in §4, we fix ℓ_0, ℓ_1, and $\ell = \ell_0 + \ell_1$. Define

$$(5.1)\quad C = K_c[u_{i,\alpha\beta}^{(m)}], \qquad m \in \mathbf{Z}_+, \qquad 1 \le \alpha, \beta \le \ell,$$

where $K_c = Ker\,(\partial|_K)$ is the subring of ∂-constants in K, and the index i in (5.1) runs over either

$$(5.2i)\quad 0 \le i < n, \qquad n \in \mathbf{N},$$

or

$$(5.2ii)\quad -\infty < i < n, \qquad n \in \mathbf{N},$$

while n is fixed, and the range of (α, β) for $i = n-1$ is defined by (5.6ii) below.

The $\mathbf{Z}_2$–grading on C is defined by

$$(5.3)\quad p\left(u_{i,\alpha\beta}^{(m)}\right) = p(\alpha) + p(\beta),$$

$$(5.4)\quad p(\alpha) = 0, \alpha \le \ell_0; \qquad p(\alpha) = 1, \ \alpha > \ell_0.$$

Now let us fix the following *even* operator

$$(5.5)\quad L = \sum^{n} u_i \xi^i \in Mat_\ell(\mathcal{O}_C), \qquad (u_i)_{\alpha\beta} := u_{i,\alpha\beta},$$

whose two highest coefficients, u_n and u_{n-1}, satisfy the conditions:

(5.6*i*) $u_n \in Mat_\ell(K_c)$ is (even) diagonal, $u_n = \mathrm{diag}\,(\mathrm{k}_1, \ldots \mathrm{k}_\ell)$, with all k_α's invertible in K_c, and with all $(\mathrm{k}_\alpha - \mathrm{k}_\beta)$'s invertible in K_c whenever $\mathrm{k}_\alpha \neq \mathrm{k}_\beta$. [Equivalently, u_n is (even) diagonalizable and invertible; $ad\,u_n$ is invertible on $Im(ad\,u_n)$ in $Mat_\ell(K_c)$ (and, hence, in $Mat_\ell(E')$ for any K_c-bimodule E').]

(5.6*ii*) $u_{n-1,\alpha\beta} = 0$ whenever $\mathrm{k}_\alpha = \mathrm{k}_\beta$; in particular, $u_{n-1,\alpha\alpha} = 0$. [Equivalently, $u_{n-1} \in Im\,(ad\,u_n)$.]

We shall always assume that L is nondegenerate, i.e., $L \neq const\,\mathbf{1}\xi$.

5.7. Definition. The centralizer of L, $Z(L)$, is defined as

$$(5.8)\quad Z(L) = \{P \in Mat_\ell(\mathcal{O}_C) | PL = LP\}.$$

5.9. Definition. A Lax equation with the Lax operator L given by (5.5), (5.6), is an equation of the form

$$(5.10)\quad \partial_P(L) = [P_+, L] = [-P_-, L], \qquad P \in Z(L)_0\,,$$

understood as an evolution derivation of C resulting by equating the powers of ξ in both parts of (5.10). Here we denote

$$(5.11)\quad \left(\sum r_j \xi^j\right)_+ = \sum_{j \geq 0} r_j \xi^j\,, \left(\sum r_j \xi^j\right)_- = \sum_{j<0} r_j \xi^j\,,\ r_j \in Mat_\ell(C),$$

the 'differential' and the 'integral' part, respectively, of a pseudo-differential operator $\sum r_j \xi^j$.

Since P commutes with L, both equations in (5.10) are clearly the same; from $\partial_P(L) = [-P_-, L]$ we see that $\partial_P(u_n) = 0$ and $\partial_P(u_{n-1}) \in Im\,(ad\,u_n)$. Hence, the equation (5.10) does indeed define an evolution derivation of C. In addition, if $L = L_+$ (i.e., if L is a differential operator itself, corresponding to the case (5.2i)), then $\partial_P(L) = [P_+, L]$ is a differential operator as well, so that the derivation ∂_P is compatible with the specialization $\left\{u^{(m)}_{i,\alpha\beta} = 0 | \forall i < 0\right\}$.

Our major goal in this section is to describe $Z(L)$. The properties of the Lax equations (5.10) will be then deduced from the properties of $Z(L)$.

5.12. Lemma. Let $C_1 = K[q_s^{(m)}], s \in S, m \in \mathbf{Z}_+$. Then

$$(5.13)\quad Ker\,(\partial|_{C_1}) = Ker\,(\partial|_K)\,(= K_c).$$

Proof. Let deg be the usual degree of a polynomial in C_1. We have to show that if $deg\,(f) \neq 0$ then $\partial\,(f) \neq 0, f \in C_1$. Let $u = q_s$ be present in f, and $\dfrac{\partial f}{\partial u^{(r+j)}} = 0, \forall j > 0, \dfrac{\partial f}{\partial u^{(r)}} \neq 0$, for some $r \in \mathbf{Z}_+$. Then $\dfrac{\partial}{\partial u^{(r+1)}}\partial(f) = \dfrac{\partial f}{\partial u^{(r)}} \neq 0$, so that $\partial(f) \neq 0$. ∎

5.14. Lemma. Let $C_2 = K[q_i^{(m)}], i \in I, m \in \mathbf{Z}_+$. Then, an operator $R = \sum^r R_j\xi^j \in Mat_\ell(\mathcal{O}_{C_2})$ commutes with $u_n\xi^n, [u_n\xi^n, R] = 0$, if and only if each R_j belongs to $Mat_\ell(K_c)$ and commutes with u_n.

Proof. Clearly, such an R commutes with $u_n\xi^n$. Conversely, if R commutes with $u_n\xi^n$ then equating to zero the ξ^{n+r} – and ξ^{n+r-1}–coefficients in $[u_n\xi^n, R] = 0$, we obtain

$$[u_n, R_r] = 0, \qquad nu_n\partial(R_r) + [u_n, R_{r-1}] = 0\,.$$

The first of these equalities shows that R_r commutes with u_n. In particular, if $\mathrm{k}_\alpha \neq \mathrm{k}_\beta$ then $R_{r,\alpha\beta} = 0$ (since $\mathrm{k}_\alpha - \mathrm{k}_\beta$ is invertible). When $\mathrm{k}_\alpha = \mathrm{k}_\beta$, the (α, β)–entry of the second equality yields $n\,\mathrm{k}_\alpha\,\partial\,(R_{r,\alpha\beta}) = 0$; hence, $R_{r,\alpha\beta} \in K_c$. This proves the Lemma for $j = r$. Now continue by induction on j. ∎

5.15. Definition. An element $p_r\xi^r \in Mat_\ell(\mathcal{O}_C)$ is called admissible if it is $\mathbf{Z}_2$–homogeneous and:

(5.16) p_r is constant (i.e., $p_r \in Mat_\ell(K_c)$) and diagonal;

(5.17) $(-1)^{p(\alpha)\eta}p_{r,\alpha\alpha} = (-1)^{p(\beta)\eta}p_{r,\beta\beta}$ whenever $\mathrm{k}_\alpha = \mathrm{k}_\beta$, where $\eta = p(p_r)$.

[Equivalently, p_r is constant and belongs to the center of the centralizer of u_n in $Mat_\ell(K_c)$.]

5.18. Theorem. Suppose $P = \sum^r p_j\xi^j \in Z(L)$. Then $p_r\xi^r$ is admissible.

Proof. Let us pick out the ξ^{n+r-z}–coefficients in the equality $LP = PL$, for $z = 0, 1, 2$. For $z = 0$ and 1 we obtain, respectively,

(5.19.0) $[u_n, p_r] = 0,$

(5.19.1) $[u_n, p_{r-1}] + [u_{n-1}, p_r] + u_n n\partial(p_r) = 0.$

Picking out the α - β matrix element from (5.19.0) for $\mathrm{k}_\alpha \neq \mathrm{k}_\beta$, we get

(5.20) $p_{r,\alpha\beta} = 0 \quad$ when $\mathrm{k}_\alpha \neq \mathrm{k}_\beta$.

Picking out the α - β matrix element from (5.19.1), for $\mathrm{k}_\alpha = \mathrm{k}_\beta$, we obtain

$$\mathrm{k}_\alpha n\partial(p_{r,\alpha\beta}) = [p_r, u_{n-1}]_{\alpha\beta} = \sum_\gamma [p_{r,\alpha\gamma}u_{n-1,\gamma\beta} - u_{n-1,\alpha\gamma}p_{r,\gamma\beta}],$$

and this is zero by (5.6*ii*) and (5.20). Thus,

(5.21) $\partial(p_{r,\alpha\beta}) = 0 \quad$ when $\mathrm{k}_\alpha = \mathrm{k}_\beta$.

Combining (5.20) and (5.21) we get

(5.22) $\partial(p_r) = 0.$

Now, for $z = 2$ we have, taking into account (5.22),

(5.19.2) $u_n n\partial(p_{r-1}) + [u_n, p_{r-2}] + [u_{n-1}, p_{r-1}] + [u_{n-2}, p_r] = p_r r\partial(u_{n-1})$.

Picking out the α - β matrix element from (5.19.1) for $k_\alpha \neq k_\beta$, we obtain, using (5.22) and (5.20):

$$(5.23)\ p_{r-1,\alpha\beta} = \frac{1}{k_\alpha - k_\beta} \sum_\gamma [p_{r,\alpha\gamma} u_{n-1,\gamma\beta} - u_{n-1,\gamma\alpha} p_{r,\gamma\beta}], \quad k_\alpha \neq k_\beta.$$

To proceed further we consider two separate cases:

(A) u_{n-2} is present; (B) u_{n-2} is absent, i.e., $L = u_1\xi + u_0$, $n = 1$.

Case (A) We pick out the $\alpha - \beta$ matrix element from (5.19.2) for $k_\alpha = k_\beta$, using (5.6*ii*) and (5.20):

$$k_\alpha n \partial(p_{r-1,\alpha\beta}) + [u_{n-1}, p_{r-1}]_{\alpha\beta} = [p_r, u_{n-2}]_{\alpha\beta} =$$

$$= \sum_\gamma [p_{r,\alpha\gamma} u_{n-2,\gamma\beta} - u_{n-2,\alpha\gamma} p_{r,\gamma\beta}]\ [\text{by } (5.20)] =$$

$$(5.24) \quad = \sum_{\gamma:k_\gamma = k_\alpha} p_{r,\alpha\gamma} u_{n-2,\gamma\beta} - \sum_{\gamma:k_\gamma = k_\beta} u_{n-2,\alpha\gamma} p_{r,\gamma\beta}, \quad \text{when } k_\alpha = k_\beta.$$

Since $p_{r-1,\alpha\beta} \in C$, the only way (5.24) can enter into the derivative of $p_{r-1,\alpha\beta}$ is by being zero:

$$(5.25) \quad \sum_{\gamma:k_\gamma = k_\alpha} p_{r,\alpha\gamma} u_{n-2,\gamma\beta} = \sum_{\gamma:k_\gamma = k_\beta} u_{n-2,\alpha\gamma} p_{r,\gamma\beta}, \quad \text{when } k_\alpha = k_\beta.$$

If all the k_α's are distinct then p_r is diagonal by (5.20), and (5.25) yields

$$(5.26) \quad p_{r,\alpha\alpha} = (-1)^{[p(\alpha)+p(\beta)]\eta} p_{r,\beta\beta}, \ \eta = p(p_r), \quad \text{when } k_\alpha = k_\beta,$$

and this is (5.17) if (5.21) is taken into account. If, on the contrary, not all the k_α's are distinct, let $\alpha \neq \beta$ be such that $k_\alpha = k_\beta$. Since $u_{n-2,\gamma\delta}$ are differentially independent variables, the only way (5.25) can hold true is to have $p_{r,\alpha\gamma}=0$ for $\gamma\neq\alpha, \mathrm{k}_\gamma=\mathrm{k}_\alpha$, and $p_{r,\gamma\beta}=0$ for $\gamma\neq\beta, \mathrm{k}_\gamma=\mathrm{k}_\beta$. Together with (5.20) this implies that p_r is diagonal, that (5.26) holds, and, thus, (5.17) is satified.

Case (B) In this case (5.19.2) becomes

$$(5.19.2')\quad u_1\partial(p_{r-1})+[u_1,p_{r-2}]+[u_0,p_{r-1}]=p_r r\partial(u_0).$$

Picking out the α - β matrix element from (5.19.2′), such that $\mathrm{k}_\alpha=\mathrm{k}_\beta$, we obtain, using (5.6*ii*) and (5.20):

$$\mathrm{k}_\alpha\partial(p_{r-1,\alpha\beta})=[p_{r-1},u_0]_{\alpha\beta}=$$

$$=\sum_\gamma[p_{r-1,\alpha\gamma}u_{0,\gamma\beta}-u_{0,\alpha\gamma}p_{r-1,\gamma\beta}]\ [\text{by (5.6}ii\text{) and (5.23)}]=$$

$$(5.27)\quad =\sum_{\gamma:\mathrm{k}_\gamma\neq\mathrm{k}_\alpha}\left[\frac{1}{\mathrm{k}_\alpha-\mathrm{k}_\gamma}(p_r u_0-u_0p_r)_{\alpha\beta}u_{0,\gamma\beta}\right]-$$
$$-\sum_{\gamma:\mathrm{k}_\gamma\neq\mathrm{k}_\beta}\left[u_{0,\alpha\gamma}\frac{1}{\mathrm{k}_\gamma-\mathrm{k}_\beta}(p_r u_0-u_0p_r)_{\alpha\beta}\right].$$

Since $u_0\in Im\,(adu_1)$, we can introduce another matrix $v\in Mat_\ell(C)$:

$$(5.28)\quad u_{0,\alpha\beta}=(\mathrm{k}_\alpha-\mathrm{k}_\beta)v_{\alpha\beta},\qquad v_{\alpha\beta}=0\quad\text{when }\ \mathrm{k}_\alpha=\mathrm{k}_\beta.$$

Then (5.27) becomes, by denoting $u=u_0, R=p_r$:

$$(5.29)\quad \mathrm{k}_\alpha\partial(p_{r-1,\alpha\beta})=[(uR-Ru)v+v(Ru-uR)]_{\alpha\beta}=-[Ruv+vuR]_{\alpha\beta},$$

since

$$(5.30)\quad (uRv+vRu)_{\alpha\beta}=0\quad\text{when }\ \mathrm{k}_\alpha=\mathrm{k}_\beta.$$

The latter identity can be seen as follows:

$(uRv)_{\alpha\beta}=\sum u_{\alpha x}R_{xy}v_{y\beta}=\sum(\mathrm{k}_\alpha-\mathrm{k}_x)v_{\alpha x}R_{xy}v_{y\beta}$ [by (5.20) and since $\mathrm{k}_\alpha=\mathrm{k}_\beta$] $=\sum(\mathrm{k}_\beta-\mathrm{k}_y)v_{\alpha x}R_{xy}v_{y\beta}=-(vRu)_{\alpha\beta}$.

Since $p_{r-1,\alpha\beta} \in C$, (5.29) means that $\partial(p_{r-1,\alpha\beta}) = 0$ when $\mathrm{k}_\alpha = \mathrm{k}_\beta$. Therefore, (5.29) implies

$$(5.31)\quad (Ruv + vuR)_{\alpha\beta} = 0 \quad \text{when } \mathrm{k}_\alpha = \mathrm{k}_\beta,$$

or

$$(5.32)\quad \sum_{x,y} (\mathrm{k}_x - \mathrm{k}_y)[R_{\alpha x} v_{xy} v_{y\beta} + v_{\alpha x} v_{xy} R_{y\beta}] = 0.$$

We consider two separate subcases of the case (B):

Subcase (B1) All the k_α's are different. Then p_r is constant diagonal by (5.20) and (5.21), and hence (5.17) is satisfied;

Subcase (B2) If all the k_α's were equal to each other then u_0 would vanish, and L would be degenerate. So, there is s such that $\mathrm{k}_\alpha = \mathrm{k}_\beta \neq \mathrm{k}_s, \alpha \neq \beta$. Let us pick from (5.32) all the terms containing $v_{s\beta}$:

$$(5.33)\quad \sum_x (\mathrm{k}_x - \mathrm{k}_s) R_{\alpha x} v_{xs} v_{s\beta} = (\mathrm{k}_\beta - \mathrm{k}_s) v_{\alpha s} v_{s\beta} R_{\beta\beta}.$$

It follows that $R_{\alpha x} = 0$ unless $x = \alpha$, that is, $R = p_r$ is diagonal. Furthermore, (5.33) then reduces to

$$R_{\alpha\alpha} v_{\alpha s} v_{s\beta} = v_{\alpha s} v_{s\beta} R_{\beta\beta},$$

which means

$$R_{\alpha\alpha} = (-1)^{[p(\alpha)+p(\beta)]p(R)}\, R_{\beta\beta},$$

and this is (5.17).

The following grading on C, $\mathcal{O}_C$, and $Mat_\ell(\mathcal{O}_C)$ will be very handy later on.

5.34. Definition. The w-grading on C, $\mathcal{O}_C$, and $Mat_\ell(\mathcal{O}_C)$ is defined by

$$w\left(u_{i,\alpha\beta}^{(m)}\right) = m+n-i,\ w(\xi) = w(\partial) = 1,\ w(K_c) = 0,$$

(5.35) $w(U) = w(U_{\alpha\beta})$ for $U \in Mat_\ell(\mathcal{O}_C)$.

Thus, L is w-homogeneous of w-degree n, and since multiplication of operators in $Mat_\ell(\mathcal{O}_C)$ is w-homogeneous, we can split $Z(L)$ into its w-homogeneous components.

The whole theory of superintegrable systems hinges on the following description of $Z(L)$:

5.36. Theorem. (*i*) Let $p_r\xi^r$ be an admissible element. Then there exists an unique w- and $\mathbf{Z}_2$-homogeneous element P in $Z(L)$ such that $P = p_r\xi^r +$ (lower order in ξ terms); (*ii*) $Z(L)$ consists precisely of (in general, infinite) sums of such P's; (*iii*) $Z(L)$ is an abelian Lie superalgebra, that is, every two elements that commute with L supercommute between themselves; (i^4) $Z(L)$ is generated over K_c by $Z(L)_0$.

5.37. Remark. Only the part (*i*) of this Theorem is difficult. In fact, the only available *direct* ('fractional powers') method (due to Manin [Man]) to prove this part is known only for the case $n \geq \ell$ (see Remark 6(1) in [W 1]). The only *indirect* ('dressing') method known (due to Wilson [W 1]), requires some preparation, but the main idea is this: If we find an operator $\Lambda = \mathbf{1}+\sum_{j>0}\chi_j\xi^{-j}$ which conjugates L into its constant part: $\Lambda^{-1}L\Lambda = u_n\xi^n$, then $Z(L) \subset \Lambda Z(u_n\xi^n)\Lambda^{-1}$, and $Z(u_n\xi^n)$ has been computed in Lemma 5.14. Since Λ *does not* belong to $Mat_\ell(\mathcal{O}_C)$ (this will be seen in a moment), the main problem is to find which homogeneous elements of $Z(u_n\xi^n)$ belong, after conjugation by Λ^{-1}, to $Mat_\ell(\mathcal{O}_C)$. It will turn out that these elements are precisely the admissible ones.

Let us see where the coefficients χ_j's of Λ belong to. Writing in longhand the equality $L\Lambda = \Lambda u_n \xi^n$, we obtain

$$[u_n, \chi_1] = -u_{n-1}$$

$$[u_n, \chi_2] + u_{n-1}\chi_1 + nu_n\partial(\chi_1) = -u_{n-2}$$

(5.38) $\quad$ ------------------------------------

$$[u_n, \chi_r] + u_{n-1}\chi_{r-1} + nu_n\partial(\chi_{r-1}) = \lambda_r, \quad r \geq 2,$$

where λ_r stands for an expression which does not involve χ_j's with $j \geq r-1$.

From the first equation we obtain $\chi_{1,\alpha\beta}$'s when $k_\alpha \neq k_\beta$. For $k_\alpha = k_\beta$, from the second equation we find $\partial(\chi_{1,\alpha\beta})$, since $(u_{n-1}\chi_1)_{\alpha\beta} = = \sum_{\gamma: k_\gamma \neq k_\beta} u_{n-1,\alpha\gamma}\chi_{1,\gamma\beta}$ involves only $\chi_{1,\mu\nu}$ already determined. And the same argument applies by induction on r. Thus we need to be able to 'integrate' in C, that is, to define ∂^{-1} even on those elements in C that do not belong to $Im(\partial|_C)$. Then our Λ will be constructed.

We now turn to the construction of an extension $\overline{C} \supset C$ where ∂ acts surjectively (and where, therefore, we can 'integrate' anything we want to).

Let T be a commutative superalgebra with polygradings $w_s : T \to \mathcal{L}_s, s \in S$, where $\mathcal{L}_s$ is an abelian group. Let $\mathcal{A} = \mathcal{A}(T;Q) = T[Q_j], j \in J$, be a polynomial commutative superalgebra, whose gradings are defined by elements $p(Q_j) \in \mathbf{Z}_2$, $w_s(Q_j) \in \mathcal{L}_s$. Let $\mathcal{B}$: $\mathcal{A}(T;Q) \to \mathcal{A}(T;Q)$ be a homogeneous (in all gradings) derivation maping T into itself and acting on the generators of $\mathcal{A}$ as

(5.39) $\mathcal{B}(Q_j) = Q_{b(j)}$,

with some map $b : J \to J$ satisfying the following properties:

(5.39.1) b is injective,

(5.39.2) $b^r(j) \neq j, \quad \forall j \in J, \quad \forall r \in \mathbf{N}$.

We shall need the following two Lemmas.

5.40. Lemma. $Ker\, \mathcal{B}|_{\mathcal{A}} = Ker\, \mathcal{B}|_T$.

Proof. For $f \in \mathcal{A}$, denote $\Gamma_f = \left\{ j \in J | \frac{\delta f}{\delta Q_j} \neq 0 \right\}$. If $\Gamma_f = \phi$ then $f \in T$. Suppose that $f \notin T$ so that Γ_f is nonempty. We shall show that $\mathcal{B}(f) \neq 0$. First, notice that $b(\Gamma_f)\backslash\Gamma_f \neq \phi$. Indeed, otherwise b is a permutation of Γ_f by (5.39.1); hence, some power of b is an identity which contradicts (5.39.2).

Now, pick $j \in \Gamma_f$ such that $b(j) \notin \Gamma_f$. Then $\frac{\partial}{\partial Q_{b(j)}} \mathcal{B}(f) = \frac{\partial f}{\partial Q_j} \neq 0$. ∎

5.41. Lemma. Suppose that $\mathcal{B}$ is a homogeneous derivation of $\mathcal{A}' = T[P_i], i \in I$, acting on the generators of $\mathcal{A}'$ by the rule

(5.42.1) $\mathcal{B}(P_i) = t_i \in T$,

such that

(5.42.2) t_i's are linearly independent over $Ker\, \mathcal{B}|_T$.

Then $Ker\, \mathcal{B}|_{\mathcal{A}'} = Ker\, \mathcal{B}|_T$.

Proof. Denote by deg the P-degree of elements in $\mathcal{A}'$. We shall use induction on $\deg(f), f \in \mathcal{A}'$. If $\deg(f) = 1$, say, $f = \sum a_i P_i$, $a_i \in T$, then $\mathcal{B}(f) = \sum \mathcal{B}(a_i)P_i + \sum a_i t_i$, so that $\mathcal{B}(f) = 0$ iff

(5.43.1) $\mathcal{B}(a_i) = 0$,

(5.43.2) $\sum a_i t_i = 0$.

But (5.43.1) means that $a_i \in Ker\mathcal{B}|_T$, and (5.43.2) implies, by (5.42.2), that all the a_i's vanish. More generally if $\deg(f) > 0$, choose $P = P_i$ such that $\frac{\partial f}{\partial P} \neq 0$. Say, $f = P^n a_n + P^{n-1} a_{n-1} + \dots$. If $\mathcal{B}(f) = 0$ then $\mathcal{B}(a_n) = 0$ and picking out the P^{n-1}-terms in the equality $\mathcal{B}(f) = 0$, we obtain $0 = n\mathcal{B}(P)a_n + \mathcal{B}(a_{n-1}) = \mathcal{B}(nPa_n + a_{n-1})$ and we may use induction procedure unless $\deg(nPa_n + a_{n-1}) = \deg(f)$, that is, $n = 1$. But either we can find a P entering f with degree $n \geq 2$, and we are done, or else we can't and the above argument, showing that $\mathcal{B}(a_1) = 0$ allows us to use the induction anyway since $\deg(a_1) = \deg(f) - 1$. ∎

Assume now that $\mathcal{B}$ acts trivially on T. Since $\mathcal{B}$ is homogeneous, we can split $\mathcal{A} = \mathcal{A}(T; Q)$ homogeneously as, say,

(5.44) $\mathcal{A}(T; Q) = Im\, \mathcal{B} \oplus d^0$.

Let $\{X_x^{\mathcal{A}}\}$ be a homogeneous basis over T of d^0. We introduce new variables $\{Y_x^{\mathcal{A}}\}$ and set:

(5.44i) $\mathcal{A}_1 = \mathcal{A}(T; Q, Y) = \mathcal{A}[Y_x^{\mathcal{A}}] = T[Q_j, Y_x^{\mathcal{A}}]$,

(5.44ii) $\mathcal{B}_1(Y_x^{\mathcal{A}}) = X_x^{\mathcal{A}}, \mathcal{B}_1|_{\mathcal{A}} = \mathcal{B},\ \mathcal{B}_1 \in Der\,(\mathcal{A}_1)$,

(5.44iii) $\mathrm{w}(Y_x^{\mathcal{A}}) = \mathrm{w}(X_x^{\mathcal{A}}) - \mathrm{w}(\mathcal{B})$, where w stands for each one of the gradings. Thus, $\mathcal{B}_1$ is homogeneous, with all its gradings being the same as those of $\mathcal{B}$. Also, $Im\,\mathcal{B}_1 \supset \mathcal{A}$, and $Ker\,\mathcal{B}_1 = Ker\,\mathcal{B} = T$. Indeed, by Lemma 5.40, $Ker\,\mathcal{B}|_{\mathcal{A}} = Ker\,\mathcal{B}|_T$. Hence, by Lemma 5.41, $Ker\,\mathcal{B}_1|_{\mathcal{A}_1} = Ker\,\mathcal{B}_1|_{\mathcal{A}} = Ker\,\mathcal{B}|_{\mathcal{A}} = T$ as well.

The net result is that we have constructed an extension $(\mathcal{A},\mathcal{B}) \subset (\mathcal{A}_1,\mathcal{B}_1)$ such that $\mathcal{A} \subset Im\,\mathcal{B}_1, \mathrm{w}(\mathcal{B}_1) = \mathrm{w}(\mathcal{B})$, and $Ker\,\mathcal{B}_1 = Ker\,\mathcal{B} = T$. Continuing in the same fashion we obtain an extension $(\overline{\mathcal{A}},\overline{\mathcal{B}}) \supset (\mathcal{A},\mathcal{B})$, by simply taking $\overline{\mathcal{A}} = \cup_{j\geq 0}\mathcal{A}_j$, with a homogeneous derivation $\overline{\mathcal{B}}$ on $\overline{\mathcal{A}}$ being an epimorphism, $\overline{\mathcal{B}}|_{\mathcal{A}_i} = \mathcal{B}_i$, and with $Ker\,\overline{\mathcal{B}} = Ker\,\mathcal{B} = T, \mathrm{w}(\overline{\mathcal{B}}) = \mathrm{w}(B)$.

Applying this construction to the case $T = K_c, \{Q_j\} = \{u^{(m)}_{i,\alpha\beta}\}, \mathcal{A} = C$, $\mathcal{B} = \partial$, and $w_1 = w$ given by (5.35), we obtain the following result:

5.45. Theorem. There exists and unique an even operator Λ in $Mat_\ell(\mathcal{O}_{\overline{C}})$, of w-degree zero, and of the form $\Lambda = \mathbf{1} + \sum\limits_{j>0} \chi_j \xi^{-j}, \chi_j \in Mat_\ell(\overline{C})$, such that $\Lambda^{-1} L\,\Lambda = u_n \xi^n$.

Proof. In analysing the equations (5.38) we have seen already that Λ exists provided we can 'integrate' in $\overline{C}$, and we have just gotten ourselves this privilege. Since all u_i's in (5.38) are even, the resulting χ_j's will be even too. Furthermore, since $Ker\overline{\partial} = Ker\partial = K_c$ has w-grading zero, each time we find χ_r by 'integrating' we can fix the 'constant of integration' to be zero, since $w(\Lambda) = 0$ requires $w(\chi_j) = j > 0$, and our equations (5.38) are w-homogeneous as components of the w-homogeneous equation $L\Lambda = \Lambda u_n \xi^n$. ∎

We need one more 'preparational' result about the extension $\overline{\mathcal{A}} \supset \mathcal{A}$.

5.46. Lemma. Let Z: $\mathcal{A}_j \to \mathcal{A}_j$ be a homogeneous derivation over T that supercommutes with $\mathcal{B}_j$. Then it can be extended to a homogeneous derivation of $\overline{\mathcal{A}}$ that supercommutes with $\overline{\mathcal{B}}$.

Proof. It is enough to extend Z homogeniously into $\mathcal{A}_{j+1}$, such that $[Z, \mathcal{B}_{j+1}] = 0$, and then to iterate the procedure. To do that, we have to apply $[Z, \mathcal{B}_{j+1}]$ to every new homogeneous generator $Y \in \{Y_x^{\mathcal{A}_j}\}$ of $\mathcal{A}_{j+1}$, and to make sure that we get zero:

$$(5.47)\quad \mathcal{B}_{j+1}Z(Y) = (-1)^{p(\mathcal{B})p(Z)} Z\mathcal{B}_{j+1}(Y).$$

But $\mathcal{B}_{j+1}(Y) = X \in \mathcal{A}_j$, and $\mathcal{A}_j \in Im\mathcal{B}_{j+1}$. Hence, we can find $Z(Y)$ from (5.47), and since $\mathcal{B}_{j+1}$ and Z on $\mathcal{A}_j$ are homogeneous, we can make $Z(Y)$ and, thus, Z on $\mathcal{A}_{j+1}$, homogeneous as well. ∎

5.48. Remark. Since $Ker\, \mathcal{B}_{j+1} = T$, we can extend Z on $\overline{\mathcal{A}}$ uniquely provided the gradings of T will not match those of $Z(Y)$ in the equation (5.47), for all $Y \in \{Y_x^{\mathcal{A}_i}\}, i \geq j$.

Proof of Theorem 5.36. By Lemma 5.14 and Definition 5.15, an element $p_r\xi^r$ is admissible if and only if it belongs to the center of the centralizer of $u_n\xi^n$ in $Mat_\ell(\mathcal{O}_{\overline{C}})$. Conjugating by Λ, we see that an element $P = p_r\xi^r +$ (lower order in ξ terms) from $Mat_\ell(\mathcal{O}_{\overline{C}})$ starts with an admissible element if and only if P belongs to the center of the centralizer of L in $Mat_\ell(\mathcal{O}_{\overline{C}})$, so that everything which commutes with L supercommutes with P as well.

To prove (i) we set $P = \Lambda p_r\xi^r\Lambda^{-1} = \sum_{-\infty}^{r} p_j\xi^j$ and suppose that $P \notin Mat_\ell(\mathcal{O}_C)$. Say, $p_s \notin Mat_\ell(C)$. We arrive at a contradiction as follows. Let $\mathcal{A}_m$ be the smallest among the algebras $\mathcal{A}_j$ containing all the entries of p_s, and let $Y \in \{Y_x^{\mathcal{A}_{m-1}}\}$ be one of the new generators in $\mathcal{A}_m$ that one of the matrix elements in p_s actually depends upon. Let us deduce that $\dfrac{\partial}{\partial Y}(p_s) = 0$.

It is obvious that $\dfrac{\partial}{\partial Y}$ commutes with $\mathcal{B}_m$ on $\mathcal{A}_m$. Hence, by Lemma 5.46, it can be extended to act on $\overline{\mathcal{A}}$ commuting with $\overline{\mathcal{B}}$. Denote by the symbol ∂_Y both this extension and also the extension of $\frac{\partial}{\partial Y}$ to $Mat_\ell(\mathcal{O}_{\overline{C}})$ according to Definition 5.62 and Remark 5.68 below. Then, by (5.61) below, we have

$$(5.49)\quad \partial_Y(L) = [\partial_Y(\Lambda)\Lambda^{-1}, L],$$

$$(5.50)\quad \partial_Y(P) = [\partial_Y(\Lambda)\Lambda^{-1}, P].$$

Since $\partial_Y(L) = 0$ it follows from (5.49) that $\partial_Y(\Lambda)\Lambda^{-1}$ commutes with L. Hence, it commutes with P. Therefore, by (5.50), $\partial_Y(P) = 0$. In particular, $\partial_Y(p_s) = 0$. This proves (i).

The (ii) part follows from the (i) part and Theorem (5.18).

The (iii) part follows from the (i) part and the fact that $Z(L)$ in $Mat_\ell(\mathcal{O}_C)$ is the center of $\{Z(L) \text{ in } Mat_\ell(\mathcal{O}_{\overline{C}})\}$.

The (i^4) part follows from Proposition 5.58 below. ∎

It remains to tie up a few loose ends we have met in the above Proof.

5.51. Definition. Let K be a commutative superalgebra. For an element $\mathrm{k} \in K$, let $\hat{\mathrm{k}} \in Mat_\ell(K)$ be the following diagonal matrix:

$$(5.52)\quad \hat{\mathrm{k}}_{\alpha\beta} = \delta^\alpha_\beta(-1)^{p(\alpha)p(\mathrm{k})}\mathrm{k}.$$

5.53. Lemma. For $\mathrm{k}, \mathrm{k}_1 \in K$ and $A \in Mat_\ell(E)$, where E is a $\mathbf{Z}_2$-graded K–bimodule,

$$(5.54)\quad \hat{\mathrm{k}}A = (-1)^{p(\mathrm{k})p(A)}A\hat{\mathrm{k}},$$

$$(5.55)\quad \hat{\mathrm{k}}_1\hat{\mathrm{k}}A = \widehat{\mathrm{k}_1\mathrm{k}}A.$$

Proof. We have,

$$(\hat{\mathrm{k}}A)_{\alpha\beta} = (-1)^{p(\mathrm{k})p(\alpha)}\mathrm{k}A_{\alpha\beta} = (-1)^{p(\mathrm{k})p(\alpha)}A_{\alpha\beta}\mathrm{k}(-1)^{p(\mathrm{k})[p(A)+p(\alpha)+p(\beta)]} =$$
$$= (-1)^{p(\mathrm{k})p(A)}A_{\alpha\beta}\mathrm{k}(-1)^{p(\beta)p(\mathrm{k})} = (-1)^{p(\mathrm{k})p(A)}(A\hat{\mathrm{k}})_{\alpha\beta},$$

which proves (5.54). Now,

$$(\hat{\mathrm{k}}_1\hat{\mathrm{k}}A)_{\alpha\beta} = (-1)^{p(\alpha)p(\mathrm{k}_1)}\mathrm{k}_1(-1)^{p(\alpha)p(\mathrm{k})}\mathrm{k}A_{\alpha\beta} =$$
$$= (-1)^{p(\alpha)[p(\mathrm{k}_1)+p(\mathrm{k})]}\mathrm{k}_1\mathrm{k}A_{\alpha\beta} = (\widehat{\mathrm{k}_1\mathrm{k}}A)_{\alpha\beta},$$

which proves (5.55). ∎

5.56. Proposition. If $A, B \in Mat_\ell(\mathcal{O}_C)$ and $[A, B] = 0$ then $[\hat{\mathrm{k}}A, B] = [A\hat{\mathrm{k}}, B] = 0, \forall \mathrm{k} \in K_c$.

Proof. The second equality follows from the first one and (5.54). Now,
$[\hat{\mathrm{k}}A, B] = \hat{\mathrm{k}}AB - (-1)^{[p(\mathrm{k})+p(A)]p(B)}B\hat{\mathrm{k}}A$ [by (5.54)]=
$= \hat{\mathrm{k}}AB - (-1)^{p(A)p(B)}\,\hat{\mathrm{k}}BA = \hat{\mathrm{k}}[A, B] = 0.$ ∎

5.57. Corollary. $Z(L)$ is a K_c-bimodule.

5.58. Proposition. The center of $Z(u_n\xi^n)$ is generated, over K_c, by admissible elements $p_r\xi^r$ where p_r has only 0's and 1's on the diagonal.

Proof. By (5.17), $p(p_r) = p(p_{r,\alpha\alpha})$. Hence, $p_r = \sum \widehat{p_{r,\alpha\alpha}}E_\alpha$, where E_α is an admissible matrix with $E_{\alpha,\beta\beta} = 1$ or 0 depending upon whether k_β equals k_α or not, respectively. ∎

5.59. Lemma. Let E be an associative $\mathbf{Z}_2$-graded algebra, Λ and R be homogeneous elements in E, and Λ^{-1} be the inverse of Λ. If $Z : E \to E$ is a homogeneous derivation, then

(5.60) $Z(\Lambda^{-1}) = -(-1)^{p(Z)p(\Lambda)}\Lambda^{-1}Z(\Lambda)\Lambda^{-1}$,

and

(5.61) $\{Z(\Lambda^{-1}R\Lambda) = 0\} \Longrightarrow \{Z(R) = [Z(\Lambda)\Lambda^{-1}, R]\}$.

Proof. From $\Lambda\Lambda^{-1} = 1$ we see that $p(\Lambda^{-1}) = p(\Lambda)$. Now,
$0 = Z(\Lambda^{-1}\Lambda) = Z(\Lambda^{-1})\Lambda + (-1)^{p(Z)p(\Lambda)}\Lambda^{-1}\, Z(\Lambda)$,
and (5.60) follows. Also, writing in long hand the equality $Z(\Lambda^{-1}R\Lambda) = 0$ and using (5.60), we obtain

$$0 = -(-1)^{p(Z)p(\Lambda)}\Lambda^{-1}Z(\Lambda)\Lambda^{-1}R\Lambda + (-1)^{p(Z)p(\Lambda)}\Lambda^{-1}Z(R)\Lambda + \\ + (-1)^{p(Z)[p(\Lambda)+p(R)]}\Lambda^{-1}RZ(\Lambda),$$

and (5.61) follows. ∎

5.62. Definition. Let E be an associative $\mathbf{Z}_2$–graded algebra, E' a E–bimodule, and Z: $E \to E'$ a $\mathbf{Z}_2$–homogeneous derivation. This derivation extends to a $\mathbf{Z}_2$–homogeneous derivation Z:$Mat_\ell(E) \to Mat_\ell(E')$, of the same $\mathbf{Z}_2$–grading, by the formula

$$(5.63)\quad [Z(A)]_{\alpha\beta} = (-1)^{p(\alpha)p(Z)}Z(A_{\alpha\beta}).$$

5.64. Remarks. (*i*) When $E = C$ (or $\mathcal{O}_C$), $E' = \Omega^1(C)$ (or $\mathcal{O}_{\Omega^1(C)}$), $Z = d$, we get back the Definition 4.50 ; (*ii*) In (5.63) and in Lemma 5.65 below, A does not have to be a square matrix; (*iii*) If Z is even, one gets back the naive matrix elements-wise action of derivations on matrices.

5.65. Lemma. For $\mathrm{k} \in E$ and $A, R \in Mat_\ell(E)$, one has

$$(5.66)\quad Z(\hat{\mathrm{k}}A) = \widehat{Z(\mathrm{k})}A + (-1)^{p(\mathrm{k})p(Z)}\hat{\mathrm{k}}Z(A),$$

$$(5.67)\quad Z(AR) = Z(A)R + (-1)^{p(Z)p(A)}AZ(R).$$

Proof. We have,

$$[Z(\hat{\mathrm{k}}A)]_{\alpha\beta} = (-1)^{p(\alpha)p(Z)}Z[(\hat{\mathrm{k}}A)_{\alpha\beta}] = (-1)^{p(\alpha)p(Z)}Z[(-1)^{p(\mathrm{k})p(\alpha)}\mathrm{k}A_{\alpha\beta}] =$$

$$= (-1)^{p(\alpha)[p(\mathrm{k})+p(Z)]}\left[Z(\mathrm{k})A_{\alpha\beta} + (-1)^{p(\mathrm{k})p(Z)}\mathrm{k}Z(A_{\alpha\beta})\right] =$$

$$= \left[\widehat{Z(\mathrm{k})}A\right]_{\alpha\beta} + (-1)^{p(\mathrm{k})p(Z)}\hat{\mathrm{k}}[Z(A)]_{\alpha\beta},$$

which yields (5.66). Similarly,

$$[Z(AR)]_{\alpha\beta} = (-1)^{p(\alpha)p(Z)}Z[(AR)_{\alpha\beta}] =$$

$$= (-1)^{p(\alpha)p(Z)}\left\{\sum_{\gamma}[Z(A_{\alpha\gamma})R_{\gamma\beta} + (-1)^{p(Z)[p(A)+p(\alpha)+p(\gamma)]}A_{\alpha\gamma}Z(R_{\gamma\beta})]\right\} =$$

$$[Z(A)R]_{\alpha\beta} + (-1)^{p(Z)p(A)}[AZ(R)]_{\alpha\beta},$$

which proves (5.67). ∎

5.68. Remark. If, in Definition 5.62, G and ∂'s act on E and E' and their actions commute with Z, this Z can be extended to a derivation $Z : Mat_{\ell}(\mathcal{O}_E) \to Mat_{\ell}(\mathcal{O}_{E'})$ by the same formula (5.63). Lemma 5.65 will then remain valid.

Now that we have an exact description of $Z(L)$ at our disposal, we can readily deduce the basic properties of the classical superintegrable systems. But first, we need a bit of information about properties of derivations extended from C into $Mat_{\ell}(C)$ and $Mat_{\ell}(\mathcal{O}_C)$ as described in Definition 5.62 and Remark 5.68.

5.69. Proposition. Let $X : C \to C$ be an evolution derivation. Then its extension into $Mat_{\ell}(C)$ (or $Mat_{\ell}(\mathcal{O}_C)$) commutes with the corresponding extension of ∂.

Proof follows from (5.63). ∎

5.70. Lemma. Let Z, E, and E' be as in Definition 5.62. Then

$$(5.71)\quad str\,[Z(A)] = Z[str\,(A)], \qquad \forall A \in Mat_{\ell}(E).$$

Proof. We have, by (4.5) and (5.63),

$$str\,[Z(A)] = \sum(-1)^{p(\alpha)[1+p(Z(A))]}\,[Z(A)]_{\alpha\alpha} =$$

$$= \sum(-1)^{p(\alpha)[1+p(Z)+p(A)]}\,(-1)^{p(Z)p(\alpha)}\,Z(A_{\alpha\alpha}) =$$

$$= Z\left(\sum(-1)^{p(\alpha)[1+p(A)]}\,A_{\alpha\alpha}\right) = Z[str\,(A)].$$ ∎

5.72. Proposition. Suppose L in (5.5) has u_n already diagonal. Then the Lax equations (5.10) make sense for all $P \in Z(L)$ (and *not* only for even P's), provided ∂_P acts on L in the left-hand-side of (5.10) in the sense of Definition 5.62 (and not matrix elements-wise).

Proof. Let P be odd. We have to check that the condition (5.6*ii*) is preserved by ∂_P. We have

$$(-1)^{p(\alpha)}\partial_P(u_{n-1,\alpha\beta}) = [u_n, p_{-1}]_{\alpha\beta} = (\mathrm{k}_\alpha - \mathrm{k}_\beta)p_{-1,\alpha\beta},$$

so that

$$(5.73)\quad \partial_P(u_{n-1,\alpha\beta}) = [u_n, \tilde{p}_{-1}]_{\alpha\beta}, \quad \tilde{p}_{-1,\alpha\beta} := (-1)^{p(\alpha)}\,p_{-1,\alpha\beta}.$$ ∎

5.74. Remark. If u_n is not diagonal, and $S \in Mat_\ell(K_c)$ conjugates u_n into its diagonal form, the Proposition 5.72 holds true provided S is $\mathbf{Z}_2$-*homogeneous*. Let us agree in what follows to consider only such situations.

5.75. Lemma. Let $P, R \in Z(L)$, with P being w-homogeneous of non-negative ξ-degree (that is, if $P = p_r\xi^r$ + (lower order in ξ terms), then $w(P) = \deg_\xi(P) = r \geq 0$). Then

$$(5.76)\quad \partial_P(R) = [-P_-, R] = [P_+, R].$$

Proof. Applying ∂_P to the equality $[R, L] = 0$ and using equalities $\partial_P(L) = [-P_-, L]$ and $[P, L] = 0$, we find that

(5.77) $[\partial_P(R) - (-1)^{p(P)p(R)}[R, P_-], L] = 0.$

Thus the operator $\partial_P(R)+[P_-, R]$ commutes with L. I claim that this operator vanishes. In view of Theorem 5.36(ii) it is enough to consider the case when R is w-homogenious. Then the w-degree of our operator equals to $w(P) + w(R)$, while its ξ-degree is not more then $w(R) - 1$. By Theorem 5.35(i) the operator must be zero. This yields the first equality in (5.76). The second one then follows at once from $[P_+ + P_-, R] = 0$. ∎

5.78. Corollary. For $P, R \in Z(L)$,

(5.79) $\partial_P(R_+) = [-P_-, R_+]_+.$

Proof. We have, using (5.76),

$$\partial_P(R_+) = [\partial_P(R)]_+ = [-P_-, R]_+ = [-P_-, R_- + R_+]_+ = [-P_-, R_+]_+. \quad ∎$$

5.80. Theorem. For $P, R \in Z(L)$, the derivations ∂_P and ∂_R supercommute.

Proof. We show that $[\partial_P, \partial_R](L) = 0$. We have, using (5.10) and (5.79),

$$\partial_P\partial_R(L) = \partial_P([R_+, L]) = [[-P_-, R_+]_+, L] + (-1)^{p(P)p(R)}[R_+, [P_+, L]],$$

$$(-1)^{p(P)p(R)}\partial_R\partial_P(L) = (-1)^{p(P)p(R)}[[-R_-, P_+]_+, L] + [P_+, [R_+, L]].$$

Subtracting and using the graded Jacobi identity we find that $[\partial_P, \partial_R](L)$ equals to the commutator of L with

$$[P_-, R_+]_+ - (-1)^{p(P)p(R)}[R_-, P_+]_+ + [P_+, R_+] =$$

$$= [P_-, R_+]_+ + [P_+, R_-]_+ + [P_+, R_+] = [P_- + P_+, R_- + R_+]_+ =$$
$$= [P, R]_+ = 0,$$

the last equality since $[P, R] = 0$. ∎

5.81. Theorem. All the evolution derivations $\partial_P : C \to C, P \in Z(L)$, have a common infinite set of c.l.'s $\{str\, Res\,(R), R \in Z(L)\}$.

Proof. We have,

$$\partial_P[str\, Res\,(R)]\,[\text{by } (5.71)] = str\, \partial_P[Res\,(R)] =$$
$$= str\, Res\,[\partial_P(R)[\text{by } (5.76] = str\, Res\,([P_+, R])[\text{by } (4.40)] \sim 0.$$ ∎

5.82. Remark. In view of Propositin 5.83 below, and Proposition 5.58, nothing is gained in considering odd P's in $Z(L)$.

5.83. Proposition. Let $\mathrm{k} \in K$, $A \in Mat_\ell\,(E)$, where E is a K-module. Then

$$str\,(\hat{\mathrm{k}}A) = \mathrm{k}\, str\,(A).$$

Proof. We have,

$$str\,(\hat{\mathrm{k}}A) = \sum_\alpha (-1)^{p(\alpha)[1+p(\mathrm{k})+p(A)]}\,(-1)^{p(\alpha)p(\mathrm{k})}\mathrm{k}A_{\alpha\alpha} =$$
$$= \mathrm{k}\sum_\alpha (-1)^{p(\alpha)[1+p(A)]}\,A_{\alpha\alpha} = \mathrm{k}\, str\,(A).$$ ∎

5.84. Corollary. For $\mathrm{k} \in K$, $R \in Mat_\ell\,(\mathcal{O}_E)$,

(5.85) $str\, Res\,(\hat{\mathrm{k}}R) = \mathrm{k}\, str\, Res(R)$.

Proof is obvious. ∎

Theorems 5.36, 5.80, and 5.81 generalize for the clasical superintegrable systems the corresponding properties (1.8i–iii) of the classical integrable system. We now turn to the Hamiltonian properties of the classical superintegrable systems, generalizing the property (1.9).

§6. Variational Derivatives of Conservation Laws and the SuperHamiltonian Structure of Classical Superintegrable Systems

In this section we compute the variational derivatives of conservation laws (Theorem 6.13 and formula (6.16)), find a Hamiltonian structure of Lax equations (5.10) and associated to it Lie superalgebras (Theorems 6.21 and 6.26), describe elements in the Kernel of this Hamiltonian structure (Theorem 6.32), and examine the problem of nontriviality of *c.l.*'s (Theorem 6.34).

We use the notation of §5. Let τ be the dimension of the center $CZ(u_n)$ of the centralizer $Z(u_n)$ of u_n in $Mat_\ell(K_c)$. Define $E_\alpha \in Mat_\ell(\mathbf{Z})$, $\alpha = 1, \ldots, \tau$, by the formula

$$(6.1) \quad (E_\alpha)_{\beta\gamma} = \delta_\gamma^\beta \text{ times } \begin{cases} 1, & \text{if } k_\beta = k_\alpha \\ 0, & \text{otherwise} \end{cases}$$

In other words, E_α's form a basis of $CZ(u_n)$. We denote by $X_\alpha^{[r]}$, $r \in \mathbf{Z}$, the unique w - and $\mathbf{Z}_2$ - homogeneous element of $Z(L)$ of the form $X_\alpha^{[r]} = E_\alpha \xi^r +$ (lower order terms in ξ). In other words, $X_\alpha^{[r]} = \Lambda E_\alpha \xi^r \Lambda^{-1}$, and $Z(L)$ consists, by Theorem 5.36(*ii*), of linear combinations of $X_\alpha^{[r]}$'s with coefficients in K_c. Obviously,

$$(6.2) \quad X_\alpha^{[r]} X_\alpha^{[s]} = X_\alpha^{[r+s]}, \quad r, s \in \mathbf{Z},$$

$$(6.3) \quad X_\alpha^{[r]} X_\beta^{[s]} = 0 \text{ for } \alpha \neq \beta,$$

since the corresponding assertions hold for the Λ - conjugate elements $E_\alpha \xi^r$, etc.

In the next five Lemmas we shall write $R \equiv S$ for $R, S \in Mat_\ell(\mathcal{O}_{\Omega^1(C)})$ if $R - S$ is a sum of commutators of the form $[X_\alpha^{[r]}, w]$, $w \in Mat_\ell(\mathcal{O}_{\Omega^1(C)})$. By Theorem 4.39, $R \equiv S$ implies $str\, Res\,(R) \sim str\, Res\,(S)$.

6.4. Lemma. If $R \equiv S$ then $X_\alpha^{[r]} R \equiv X_\alpha^{[r]} S$ and $R X_\alpha^{[r]} \equiv S X_\alpha^{[r]}$.

Proof. If $[a, b] = 0$ and $p(a) = 0$ then $a[b, c] = [b, ac]$ and $[b, c]a = [b, ca]$. Now take $a = X_\alpha^{[r]}$, $[b, c] = R - S$, and use (6.2), 6.3). ∎

Denote $X_\alpha = X_\alpha^{[1]}$.

6.5. Lemma. $d(X_\alpha^{[r]}) \equiv r X_\alpha^{[r-1]} d(X_\alpha)$.

Proof. For $r \geq 1, X_\alpha^{[r]} = (X_\alpha)^r$. Thus, for $r \geq 2$,

$$d(X_\alpha^{[r]}) = d\left[(X_\alpha)^r\right] \equiv r(X_\alpha)^{r-1} d(X_\alpha) = r X_\alpha^{[r-1]} d(X_\alpha),$$

which proves Lemma for $r \geq 2$. Now apply d to the equality $X_\alpha^{[0]} = (X_\alpha^{[0]})^s$ with $s \geq 2$, resulting in $d(X_\alpha^{[0]}) \equiv s X_\alpha^{[0]} d(X_\alpha^{[0]})$. Hence $d(X_\alpha^{[0]}) \equiv 0$, which proves (6.5) for $r = 0$. For $r = 1$, we apply d to the equality $X_\alpha = X_\alpha^{[0]} X_\alpha$, and use $d(X_\alpha^{[0]}) \equiv 0$, getting $d(X_\alpha) \equiv X_\alpha^{[0]} d(X_\alpha)$.

Denote $Y_\alpha = X_\alpha^{[-1]}$, so that $X_\alpha^{[-s]} = (Y_\alpha)^s$ for $s \geq 1$. From the chain of relations $0 \equiv d(X_\alpha^{[0]}) = d(X_\alpha Y_\alpha) \equiv X_\alpha d(Y_\alpha) + Y_\alpha d(X_\alpha)$, we conclude that

(6.6) $X_\alpha^{[r]} d(Y_\alpha) \equiv -X_\alpha^{[r-2]} d(X_\alpha),\ r \in \mathbf{Z}$.

Now, for $-r \geq 2$, we have

$$\begin{aligned} d(X_\alpha^{[r]}) &= d\left[(Y_\alpha)^{-r}\right] \equiv -r(Y_\alpha)^{-r-1} d(Y_\alpha) = \\ &= -r X_\alpha^{[1+r]} d(Y_\alpha) [\text{by } (6.6)] \equiv r X_\alpha^{[r-1]} d(X_\alpha), \end{aligned}$$

which proves (6.5) for $r \leq -2$. Finally, $d(X_\alpha^{[-1]}) = d(X_\alpha^{[0]} X_\alpha^{[-1]}) \equiv$ $\equiv X_\alpha^{[0]} d(X_\alpha^{[-1]})$ [by (6.6)] $\equiv -X_\alpha^{[-2]} d(X_\alpha)$, which is (6.5) for $r = -1$. ∎

6.7. Lemma. (i) $r X_\alpha^{[r]} d(X_\alpha^{[s]}) \equiv s X_\alpha^{[s]} d(X_\alpha^{[r]})$; (ii) $X_\alpha^{[r]} d(X_\beta^{[s]}) \equiv 0$ for $\alpha \neq \beta$.

Proof. (i) From Lemma 6.5, $rX_\alpha^{[r]}d(X_\alpha^{[s]}) \equiv rsX_\alpha^{[r+s-1]}d(X_\alpha)$, and this expression is symmetric in (r,s). (ii) From the same Lemma,

$$X_\alpha^{[r]}d(X_\beta^{[s]}) \equiv sX_\alpha^{[r]}X_\beta^{[s-1]}d(X_\beta) = 0 \text{ by } (6.3). \qquad \blacksquare$$

6.8. Lemma. Let $P, Q \in Z(L)$ be even, w-homogeneous, of w-degrees r and s respectively. Then

$$(6.9)\quad rPd(Q) \equiv sQd(P).$$

Proof. Write $P = \sum p_\alpha X_\alpha^{[r]}$, $Q = \sum q_\beta X_\beta^{[s]}$, and use Lemma 6.7. ∎

6.10. Lemma. Let $P \in Z(L)$ be even, w-homogeneous, of w-degree r. Then

$$(6.11)\quad d(LP) \equiv \frac{n+r}{n} d(L)P.$$

Proof. Use (6.9) with $Q = L$, to get

$$d(LP) = d(L)P + Ld(P) \equiv d(L)P + \frac{r}{n}Pd(L) \equiv \frac{n+r}{n}d(L)P. \qquad \blacksquare$$

If $P \in Z(L)_0$ is w-homogeneous of w-degree r, we define

$$(6.12)\quad H_P = \begin{cases} r^{-1}n\, str\, Res(P) & ,r > 0, \\ 0 & ,r \leq 0. \end{cases}$$

If $P \in Z(L)_0$ is not w-homogeneous, we define H_P by adding (6.12) over the w-homogeneous components of P. By Theorem 5.81 and Remark 5.82, H_P's are *c.l.*'s of the Lax derivations (5.10).

6.13. Theorem. $d(H_{LP}) \sim str\, Res\,[d(L)P]$, $\forall P \in Z(L)_0$.

Proof. Apply *str Res* to (6.11) and use (4.57). ∎

6.14. Remark. The reader may have noticed that one could derive the $w(P) \geq 0$ –case of Theorem 6.13 simply by making use of (4.60) and (4.64). The reason we chose instead the above route (modelled on [K – W]) is to have Theorem 6.13 established also for the case $w(P) < 0$ which we shall use in the Proof of Theorem 6.32 below.

Theorem 6.13 is the starting point of the Hamiltonian formalism of super-integrable systems. We proceed as follows.

Fix $P \in Z(L)_0$ and write it in the *left* form as

$$(6.15)\quad P = \sum \xi^j p_j, \quad p_j \in Mat_\ell(C).$$

Substituting (6.15) into (6.13), we obtain

$d(H_{LP}) \sim str\, Res\,[\sum d(u_i)\xi^i \xi^j p_j] = str\,(\sum d(u_i)p_{-i-1})$ [since $d(u_i)p_{-i-1}$ is odd] $= \sum_\alpha [d(u_i)p_{-i-1}]_{\alpha\alpha} = \sum [d(u_i)]_{\alpha\gamma} p_{-i-1,\gamma\alpha}$ [by (4.51)] $=$

$= \sum (-1)^{p(\alpha)}\, d(u_{i,\alpha\gamma})\, p_{-i-1,\gamma\alpha}.$

Hence,

$$(6.16)\quad p_{-i-1,\gamma\alpha} = (-1)^{p(\alpha)} \frac{\delta H}{\delta u_{i,\alpha\gamma}}, \qquad H = H_{LP},$$

for all i's for which $u_{i,\alpha\gamma}$ is a variable (i.e., a generator in C).
From (5.10) we obtain

$$(6.17)\quad \partial_P(L_+) = [\partial_P(L)]_+ = [-P_-, L]_+ = [L_+, P_-]_+\,,$$

$$(6.18)\quad \partial_P(L_-) = [\partial_P(L)]_- = [P_+, L]_- = [P_+, L_-]_-\,.$$

We work out (6.17) first. Denote $x_i = p_{-i-1}$, $\hat{x} = \sum_{i\geq 0} \xi^{-i-1} x_i$, so that $P_- = \sum_{i\geq 0} \xi^{-i-1} x_i = \hat{x}$. Then for $i \geq 0$,

$$\partial_P(u_i) = \{\text{right } \xi^i - \text{coefficient in } [L_+, P_-]\} =$$

$$= Res\,\{[L_+, \hat{x}]\xi^{-i-1}\} = Res\left\{[L_+, \sum_{j=0}^{n-1} \xi^{-j-1}x_j]\xi^{-i-1}\right\},$$

so that

(6.19) $\partial_P(u_{i,\alpha\beta}) = Res\,\{[L_+, \hat{x}]_{\alpha\beta}\xi^{-i-1}\}, \quad i \geq 0.$

Form the R.H.S. of (6.19) we see that

(6.20) $\partial_P(u_n) = 0,\ \partial_P(u_{n-1}) = [u_n, x_0],$

which implies that, through $\hat{x}$, only those expressions $\dfrac{\delta H}{\delta u_{i,\alpha\beta}}$ are present in (6.19) for which $i \geq 0$.

6.21. Theorem. The matrix B, which maps the vector $\left\{\dfrac{\delta H}{\delta u_{i,\alpha\beta}} | i \geq 0\right\}$ into the vector $\partial_P(u_{i,\alpha\beta})$, is superHamiltonian.

Proof. In view of (6.20), we temporarily drop off the conditions (5.6) and will consider u_n and u_{n-1} as matrices of free independent variables, restricting later the thus extended matrix B onto the invariant factor–ring generated by the relations (5.6). So set $X = (X_{i,\alpha\beta})$, where X can be informally thought of as $X_{i,\alpha\beta} = \dfrac{\delta H}{\delta u_{i,\alpha\beta}}$, $i \geq 0$.

Then, by (6.16)

(6.22) $x_{i,\alpha\beta} = (-1)^{p(\beta)}X_{i,\beta\alpha},\ \hat{x}_{\alpha\beta} = \sum \xi^{-i-1}(-1)^{p(\beta)}X_{i,\beta\alpha} + \dots,$

and therefore, for arbitrary even vector $Y = (Y_{i,\alpha\beta})$, we have

$$B(X)^tY = \sum B(X)_{i,\alpha\beta}Y_{i,\alpha\beta} = Res\left\{[L_+, \hat{x}]_{\alpha\beta}\sum \xi^{-i-1}Y_{i,\alpha\beta}\right\}$$

$$= Res\left\{\sum [L_+, \hat{x}]_{\alpha\beta}\ \hat{y}_{\beta\alpha}(-1)^{p(\alpha)}\right\} = str\ Res\ \{[L_+, \hat{x}]\hat{y}\}\ [\text{by (4.44)}] \sim$$

$$\sim str\ Res\,(L_+[\hat{x}, \hat{y}]),$$

where $\hat{y}$ is gotten from Y by the same formula (6.22).

Since $[\hat{x}, \hat{y}] = -[\hat{y}, \hat{x}]$, from the expression above for $B(X)^tY$ it follows that

B is superskewsymmetric, and B is obviously even because it maps any even vector X into an even vector $Res\,\{[L_+,\hat{x}]_{\alpha\beta}\xi^{-i-1}\}$, the latter being even since the operator $[L_+,\hat{x}]$ is even due to $\hat{x}$ being even. To show that B is superHamiltonian we notice that B depends linearly upon $u_{i,\alpha\beta}$'s ; hence, by Theorem 3.82, it will be enough to show that the corresponding algebra $\mathcal{G}$ is a (stable) Lie superalgebra; it remains to check the Jacobi identity in $\mathcal{G}$.

First, we have

$$\begin{aligned}
&[B(X)^t Y] = str\,Res\,\{L_+[\hat{x},\hat{y}]\} = \\
&= str\,Res\left\{\sum u_i \xi^i [\hat{x},\hat{y}]\right\} = \\
&= str\left\{\sum u_i \text{ times left } \xi^{-i-1} \text{coefficient of } [\hat{x},\hat{y}]\right\} = \\
&= \sum (-1)^{p(\alpha)} u_{i,\alpha\beta}\left(\xi_{\text{left}}^{-i-1} \text{ of } [\hat{x},\hat{y}]_{\beta\alpha}\right),
\end{aligned}$$

so that

$$\text{(6.23)}\quad \frac{\delta}{\delta u_{i,\alpha\beta}}[B(X)^t Y] = (-1)^{p(\alpha)}\left(\xi_{\text{left}}^{-i-1} \text{ of } [\hat{x},\hat{y}]_{\beta\alpha}\right).$$

Let us denote by R the vector with components given by (6.23). If we think of X and Y as vectors of functional derivatives of linear Hamiltonians H and F respectively, then $B(X)^t Y \sim \{H,F\}$ and $R = \dfrac{\delta}{\delta \bar{u}}\{H,F\}$. Hence, by (6.22) and (6.23),

$$\begin{aligned}
\text{(6.24)}\quad \hat{r}_{\alpha\beta} &= \sum \xi^{-i-1}(-1)^{p(\beta)} R_{i,\beta\alpha} = \\
&= \sum \xi^{-i-1}(-1)^{p(\beta)}(-1)^{p(\beta)}\left(\xi_{\text{left}}^{-i-1} \text{ of} [\hat{x},\hat{y}]_{\alpha\beta}\right) = [\hat{x},\hat{y}]_{\alpha\beta},
\end{aligned}$$

so that $\hat{r} = [\hat{x},\hat{y}]$ (we discard terms of the form $\sum\limits_{i\geq n} \xi^{-i-1} f_i$ as irrelevant).

Therefore, $\mathcal{G} \approx \mathcal{G}_{<0}/\mathcal{G}_{<-n}$ where $\mathcal{G}_{<\text{k}} = \left\{\sum\limits_{i<\text{k}} f_i \xi^i | f_i \in Mat_\ell(C)\right\}^{Lie}$ ∎

6.25. Remark. In fact, we have proved more than Theorem 6.21 asserts. The superHamiltonian matrix B has turned out to be naturally associated with the Lie superalgebra $\mathcal{G}_{<0}/\mathcal{G}_{<n}$, specialized by the conditions (5.6).

We now turn to the remaining system (6.18).

6.26. Theorem. Equations (6.18) are superHamiltonian, with the Hamiltonian $H = H_{LP}$, and with the superHamiltonian matrix B whose negative is associated to the Lie superalgebra $\mathcal{G}_{\geq 0}$ of matrix differential operators with coefficients in C.

Proof. We follow the Proof of Theorem 6.21. For $i \geq 0$, we obtain,

denoting $v_i = u_{-i-1}$ and $\hat{x} = \sum_{i\geq 0} \xi^i x_i = P_+, x_{i,\alpha\beta} = (-1)^{p(\beta)} \dfrac{\delta H}{\delta v_{i,\beta\alpha}}$:

$$\partial_P(v_i) = \{\text{right } \xi^{-i-1} - \text{coefficient of } [P_+, L_-]\} =$$

$$= Res\,\{[P_+, L_-]\xi^i\} = \; Res\,\{[\hat{x}, L_-]\xi^i\},$$

so that

(6.27)

$$B(X)_{i,\alpha\beta} = \partial_P(v_{i,\alpha\beta}) = \; Res\,\{[\hat{x}, L_-]_{\alpha\beta}\xi^i\},\; \hat{x}_{\alpha\beta} = \sum_{i\geq 0} \xi^i(-1)^{p(\beta)} X_{i,\beta\alpha}.$$

Thus, the matrix B is even. Further, for an even vector Y,

$$B(X)^t Y = \sum B(X)_{i,\alpha\beta}\, Y_{i,\alpha\beta} = \; Res\,\left\{\sum [\hat{x}, L_-]_{\alpha\beta}\xi^i\, Y_{i,\alpha\beta}\right\} =$$

$$(6.28) \quad = \; Res\,\left\{\sum [\hat{x}, L_-]_{\alpha\beta}(-1)^{p(\alpha)}\, \hat{y}_{\beta\alpha}\right\} = \; str\; Res\,\{[\hat{x}, L_-]\hat{y}\} \sim$$

$$\sim -\, str\; Res\,\{L_-[\hat{x}, \hat{y}]\},$$

so that B is superskewsymmetric. Finally, from (6.28) we get

$$-B(X)^t Y \sim \; str\; Res\,\left\{\sum_{i\geq 0} v_i \xi^{-i-1}[\hat{x}, \hat{y}]\right\} =$$

$$= \; str\,\left\{\sum v_i \text{ times left } \xi^i\text{-coefficient of } [\hat{x}, \hat{y}]\right\} =$$

$$= \sum v_{i,\alpha\beta}(-1)^{p(\alpha)} \left(\xi^i_{\text{left}} \text{ of } [\hat{x},\hat{y}]_{\beta\alpha}\right),$$

which implies

$$(6.29) \quad \frac{\delta}{\delta v_{i,\alpha\beta}} [B(X)^t Y] = -(-1)^{p(\alpha)} \left(\xi^i_{\text{left}} \text{ of } [\hat{x},\hat{y}]_{\beta\alpha}\right).$$

Let R be the vector whose components are given by (6.29), $R = \frac{\delta}{\delta \bar{v}}(\{H, F\})$ for linear H and F such that $X = \frac{\delta H}{\delta \bar{v}}$, $Y = \frac{\delta F}{\delta \bar{v}}$. Then, from (6.29) we obtain

$$\hat{r}_{\alpha\beta} = \sum_{i\geq 0} \xi^i r_{i,\alpha\beta} = \sum \xi^i (-1)^{p(\beta)} R_{i,\beta,\alpha} =$$

$$= -\sum \xi^i \left(\xi^i_{\text{left}} \text{ of } [\hat{x},\hat{y}]_{\alpha\beta}\right) = -[\hat{x},\hat{y}]_{\alpha\beta},$$

so that $\hat{r} = -[\hat{x},\hat{y}]$. ∎

6.30. Remark. In the Proof of Theorem 6.26, our matrix $-B$ was associated with the differential operators from $\mathcal{G}_{\geq 0}$ considered in the *left* form. Had we wanted the matrix $-B$ to be attached to the differential operators from $\mathcal{G}_{\geq 0}$ written in the usual *right* form, we would have to work with the Lax operator L in the form

$$(6.31) \quad L = \sum_{i=0}^{n} u_i \xi^i + \sum_{i>0} \xi^{-i} u_{-i}.$$

Denote by B_+ and B_- the superHamiltonian matrices associated to the equations (6.17) and (6.18) respectively. Since B_+ (resp., B_-) depends only upon u_i's with $i \geq 0$ (resp., $i < 0$), the matrix $B = B_+ + B_-$ is also a superHamiltonian matrix. It provides us with a superHamiltonian form of the classical superintegrable systems (5.10). The following statement describes $Ker\, B$.

6.32. Theorem. If $P \in Z(L)_0$ is w-homogeneous and $0 \leq w(P) < n$ then

$str\ Res\, P \in Ker\, B$, i.e. $B\frac{\delta}{\delta \bar{u}}\,(str\ Res\ P) = 0$.

6.33. Remark. In the commutative case $l = l_0$, and for purely differential Lax operators $L = L_+$, this Theorem was given in [V] for $l = 1$ and in [K – W] for general $l \geq 1$.

Proof. By Theorems 6.13 and 6.21, $B\dfrac{\delta}{\delta \bar{u}}(str\ Res\ P) = \partial_{L^{-1}P}(L) =$ $[(L^{-1}P)_+, L] = 0$, since $w(L^{-1}P) = -n + w(P) < 0$. ∎

Naturally, for Theorem 6.32 to have any content, we have to make sure that $str\ Res\ P$ is not trivial itself. The following statement is modelled by the treatment in [K – W].

6.34. Theorem. Let $P \in Z(L)_0$ be w-homogeneous. Let $r = w(P) > 0$. (i) If r in not divisible by n then $str\ Res\ P \not\sim 0$. (ii) If r is divisible by n, say $r = n\mathrm{k}$, then: (a) if $P \neq$ const L^{k} then $str\ Res\ P \not\sim 0$; (b) if $P =$ const L^{k} then $str\ Res\ P$ is trivial or not, depending upon whether L is equal to L_+ or not.

Proof. Part $(ii$ b) is obvious.

Let first $n > 1$. Suppose $str\ Res\ P \sim 0$. From (6.16) it follows that in $(L^{-1}P)_-$ the first $(n-1)$ coefficients vanish and in the nth-coefficient few off-diagonal entries vanish corresponding to the nonvanishing entries in u_{n-1}. In particular, $Res\,(L^{-1}P) = 0$, so that $str\ Res\,(L^{-1}P) = 0$. To arrive at a contradiction, we continue the same reasoning until we hit such k that $w(L^{-\mathrm{k}}P) \leq 0$. If r is not divisible by n, then $-(n-1) \leq w(L^{-\mathrm{k}}P) < 0$. In particular, the highest term in $L^{-\mathrm{k}}P$ does not vanish. A contradiction. Suppose r is divisible by $n : r = n\mathrm{k}$, and $L^{-\mathrm{k}}P \neq \mathrm{const}\mathbf{1}$. In particular, $u_n \neq \mathrm{const}\mathbf{1}$, so that $Im\,(ad\, u_n) \neq \{0\}$ and $u_{n-1} \neq 0$. Say, $L^{-\mathrm{k}}P =: Q = q + \xi^{-1}\tilde{q} + \ldots$. Since $Q \in Z(L)$, q is even admissible. Let us see that $\tilde{q} = Res\,(L^{-\mathrm{k}}P)$ cannot vanish. Set $q = \operatorname{diag}(q_1, \ldots, q_\ell)$. Notice that not all the q_i's are equal (since $P \neq$ conts L^{k}), and also $q_\alpha = q_\beta$ whenever $\mathrm{k}_\alpha = \mathrm{k}_\beta$ in u_n (since q is admissible). Picking out the ξ^{n-1}-terms in the equality $[L, Q] = 0$ and remembering that $[u_n, q] = 0$, we obtain $[u_n, \tilde{q}] + [u_{n-1}, q] = 0$. Hence, $(\mathrm{k}_\alpha - \mathrm{k}_\beta)\tilde{q}_{\alpha\beta} = (q_\alpha - q_\beta)u_{n-1,\alpha\beta}$. Therefore, if $\tilde{q}$ is zero, u_{n-1} vanishes as well, a contradiction.

Finally, suppose $n = 1$. Denote by μ the ideal in C generated by $\{u^{(m)}_{i,\alpha\beta} | i < 0\}$. We shall prove a stronger statement: if $P \neq$ const L^r than *str Res* $P \not\equiv 0$ (mod μ). To do this we reduce L modulo μ, so that $L/(\mu)$ becomes a purely differential operator: $L/(\mu) = L_+ = u_1\xi + u_0$. Since L is nondegenerate, $u_1 \neq$ const$\mathbf{1}$ and $u_0 \neq 0$. Let $[L^{-1}P/(\mu)]_- = \xi^{-1}p + \xi^{-2}\tilde{p} + \ldots$. We have,

$$0 = [L, L^{-1}P]_-/(\mu) = [L/(\mu), L^{-1}P/(\mu)]_- = [L_+, L^{-1}P/(\mu)]_- =$$
$$= [u_1\xi + u_0, \xi^{-1}p + \xi^{-2}\tilde{p} + \ldots]_-.$$

Hence, picking out the ξ^{-1}-coefficient, we obtain

$$(6.35)\quad [u_1, \tilde{p}] + [u_0, p] + u_1\partial(p) = 0.$$

Taking the α - β matrix element of (6.35), for $\mathrm{k}_\alpha = \mathrm{k}_\beta$, we get

$$(6.36)\quad \sum_{\gamma:\mathrm{k}_\gamma \neq \mathrm{k}_\beta} (u_{0,\alpha\gamma}p_{\gamma\beta} - p_{\alpha\gamma}u_{0,\gamma\beta}) = -\mathrm{k}_\alpha\partial(p_{\alpha\beta}), \quad \mathrm{k}_\alpha = \mathrm{k}_\beta.$$

Thus, if *str Res* $P/(\mu)$ is trivial, the left-hand-side of (6.36) vanishes by (6.16). Since $w(p) = 1 + w(L^{-1}P) = w(P) > 0$, and $\mathrm{k}_\alpha \neq 0$, it follows that $p_{\alpha\beta} = 0$ whenever $\mathrm{k}_\alpha = \mathrm{k}_\beta$. Therefore, *Res* $(L^{-1}P) = 0$ (mod μ). Since taking the variational derivatives with respect to $u_{0,\alpha\beta}$ commutes with factoring out by μ, we can repeat the procedure, until we hit $Q = L^{-w(P)}P$. Then we reason as above for the case $n > 1$. ∎

CHAPTER II

LIE ALGEBRAS, KORTEWEG-DE VRIES EQUATIONS, AND BI-SUPERHAMILTONIAN SYSTEMS

§7. Three Constructions

In this section we construct three different generalizations of the Korteweg-de Vries equation: systems (7.61), (7.97), and (7.130). Each of these constructions uses a pair of generalized 2–cocycles on a differential Lie superalgebra. The proof of integrability of the constructed systems involves a complex inductive reasoning initiated in this section and completed in sections 8 and 9.

Lie algebras underlie a large part of the modern theory of continuous integrable dynamical systems. Specifically, there are two different systematically developed theories in which Lie algebras serve as a foundation: Wilson's general zero-curvature equations associated with simple complex Lie algebras [W3]; and the theory of Drinfel'd and Sokolov [D–S 1,2] of equations associated to affine Kac-Moody Lie algebras (both untwisted and twisted) (see also [F; F–K]). In both these cases the relevant Lie algebras are very far from being arbitrary and, as Drinfel'd and Sokolov demonstrate, to a large extent their theory of integrable systems is equivalent to the theory of affine Lie algebras (see also Ch. 14 in [Ka3]).

The purpose of this Chapter is to construct: 1) Two new classes of superintegrable systems (equations (7.61) and (7.97) below) associated to arbitrary metrizable Lie algebra; 2) A new class of integrable systems associated to representations of metrizable Lie algebras (equations (7.130) below). By a metrizable Lie algebra we shall understand, for a want of a better name, a finite-dimensional Lie algebra endowed with an invariant nondegenerate symmetric bilinear form on it. (Such algebras are described in [Ka 3, exerc. 2.10 and 2.11; A1–3; C-M-P; Me 1,2; Me-R 1–3]. Metrizable Lie algebras constitute next, after semisimple, most important class of Lie algebras. They can be nilpotent and solvable. As was shown in [C-M-P], if $\mathcal{G}$ is metrizable then so is the semidirect product $\mathcal{G} \underset{ad}{\propto} V(\mathcal{G})$ of $\mathcal{G}$ acting on its underlying vector space V($\mathcal{G}$) by the adjoint representation. The new metric is a linear combination of the old one (,) on $\mathcal{G}$ and $(\begin{pmatrix} X \\ x \end{pmatrix}, \begin{pmatrix} Y \\ y \end{pmatrix})_{\text{new}} = (X, y) + (x, Y)$. The latter metric can be interpreted as the linearization of the old one (,), while $\mathcal{G} \underset{ad}{\propto} V(\mathcal{G})$ can be thought of as the lineariaztion of $\mathcal{G}$. (See [Ku 17].) See also Appendix.)

To set the scene for our discussion let us recall some basic facts about the Korteweg-de Vries (KdV) equation

$$(7.1) \quad u_t = 6uu_x - u_{xxx} \,,$$

where, just for the moment. u can be considered as a function of x and t. with subscript notation in (7.1) for partial derivatives. - we shall be less informal later on. We shall pay attention to the following remarkable properties of the KdV equation (see [K-W]):

(7.2i) The equation has an infinite sequence of conserved densities $H_0 = u, H_1 = u^2, H_2 = 2u^3 + u_x^2, \ldots$, which are differential polynomials in u. i.e., polynomials (in this case over $\mathbf{Q}$) in u and its x-derivatives $u^{(j)} = (\partial/\partial x)^j(u)$;

(7.2ii) There exists an infinite number of evolution equations $u_t = X_r(u), r \in \mathbf{Z}_+$, commuting with (7.1) and between themselves: they are called the higher KdV equations:

(7.2iii) The KdV equation is a bi-Hamiltonian system: it can be written in the form

$$(7.3) \quad u_t = 6uu_x - u_{xxx} = \frac{1}{2}\partial\left(\frac{\delta H_2}{\delta u}\right) = \left(u\partial + \partial u - \frac{1}{2}\partial^3\right)\left(\frac{\delta H_1}{\delta u}\right),$$

where H_1 and H_2 are taken from (7.2i), and $\partial = \partial/\partial x$;

(7.2i^4) The conserved densities $\{H_r\}$ are in involution with respect to each of the two Hamiltonian structures featuring in (7.3);

(7.2i^5) For every $r > 0$, the rth KdV equation in (7.2ii) is also a bi-Hamiltonian system: it can be written in the form

$$(7.4) \quad u_t = B^I\left(\frac{\delta H_r}{\delta u}\right) = B^{II}\left(\frac{\delta H_{r-1}}{\delta u}\right),$$

$$(7.5) \quad B^I = \frac{1}{2}\partial, \quad B^{II} = u\partial + \partial u - \frac{1}{2}\partial^3.$$

Naturally, the properties (7.2i–i^5) are not independent: e.g., both (7.2ii) and (7.2i^4) obviously follow from (7.2i^5). (See, e.g., Lemma 9.49 below.) Also, the properties (7.2) are proven in the classical theory either directly, by using the Lax representation ([Man;W1]; or see Ch. I).

$$(7.6) \quad L_t = [P_+, L],$$

$$(7.7) \quad L = \xi^2 - u, P = 4(L^{1/2})^{2r+1}, \xi = \partial/\partial x, \quad r \in \mathbf{Z}_+,$$

or else by treating the KdV equation as a particular case of the general theory ([D-S 1,2]) corresponding to the affine Lie algebra $A_1^{(1)}$; in the former case, one has for the conserved densities H_r in (7.2i) the following simple formula ([G-Di])

$$(7.8) \quad H_r = \frac{1}{2r+1} Res(L^{1/2})^{2r+1},$$

where

$$(7.9) \quad Res\left(\sum a_i\xi^i\right) = a_{-1},$$

and one deduces $(7.2i^5)$ via the canonical Miura map ([K-W])

(7.10) $u = v^2 + v_x$,

where v satisfies the rth modified KdV (mKdV) equation

(7.11) $u_t = b\frac{\delta}{\delta v}\left(H_{r-1}\big|_{u=v^2+v_x}\right), \quad b = \frac{1}{2}\partial.$

Thus, the point of view of the classical theory on the nature of the second Hamiltonian structure B^{II} of the KdV hierarchy is this: B^{II} is the restriction of a simple Hamiltonian structure (b in (7.11)) on the Image of the Miura map.

However, the second Hamiltonian structure

(7.12) $B^{II} = u\partial + \partial u - \frac{1}{2}\partial^3$,

and the bi-Hamiltonian property $(7.2i^5)$ may be given a very different interpretation. Let K be a commutative ring with a derivation $\partial: K \to K$. Denote by $D = D(K)$ the following differential Lie algebra structure on K

(7.13) $[X, Y] = X\partial(Y) - Y\partial(X)$;

as its notation suggests, $D(K)$ can be thought of as the Lie algebra of the derivations of K of the form $K\partial$: $[X\partial, Y\partial]: = X\partial Y\partial - Y\partial X\partial = [X, Y]\partial$. Consider the following bilinear form ω on D:

(7.14) $\omega(X, Y) = cX\partial^3(Y), \quad c \in K_c = \text{ Ker } \partial\big|_K$.

Obviously, ω is a (generalized) 2-cocycle on D (see Definition 3.108), that is, ω is skewsymmetric:

(7.15) $\omega(X, Y) \sim -\omega(Y, X)$,

and

$$(7.16)\quad \omega([X,Y],Z) + \text{ c.p. } \sim 0,$$

where, as in Chapter I, $a \sim b$ means $(a-b) \in Im\partial$, and "c.p." stands for "cyclic permutation". Let $C = C_u = K[u^{(j)}], j \in \mathbf{Z}_+$, be a differential ring, with a derivation ∂ acting on the generators of C by the usual rule $\partial(u^{(j)}) = u^{(j+1)}$. Let $B = B(D)$ be the Hamiltonian operator in C associated to the Lie algebra D: B is defined by the relation (see (3.81); in this section, though, we change the sign of B to obtain more aesthetically appealing formulae):

$$(7.17)\quad B(X)Y \sim -u[X,Y],$$

so that

$$(7.18)\quad B = u\partial + \partial u.$$

Let b_ω be the operator associated to the 2-cocycle ω via the rule (see (3.100))

$$(7.19)\quad b_\omega(X)Y \sim \omega(X,Y),$$

so that

$$(7.20)\quad b_\omega = -c\,\partial^3.$$

Then the matrix $\widetilde{B} = B(D) + b_\omega$ is Hamiltonian (Theorem 3.110). In particular, taking $c = \frac{1}{2}$ in (7.14), (7.20), we find that the Hamiltonian matrix B^{II} in (7.12) is associated to the Lie algebra D and the 2-cocycle $\omega|_{c=1/2}$ (7.14) on it. The Hamiltonian matrix B^I:

$$(7.21)\quad B^I = \frac{1}{2}\partial,$$

also has an interpretation in terms of the Lie algebra D. Namely, $B^I = b_\nu$ where ν is another (trivial) 2-cocycle on D:

$$(7.22)\quad \nu(X,Y) = \frac{1}{2}\partial(X)Y.$$

[For a differential, or differential-difference, Lie algebra (or superalgebra) $\mathcal{H}$ over K, setting up the complex of skewsymmetric modulo $Im\partial$ K-valued forms on $\mathcal{H}$ (over K_c, with coefficients in the trivial representation), one arrives at the variational cohomology $H^*_{\text{var}}(\mathcal{H})$ of $\mathcal{H}$. In particular, $\dim_{K_c} H^2_{\text{var}}(D) = 1$, and $H^2_{\text{var}}(D)$ is generated over K_c by the 2-cocycle $(7.14)|_{c=1}$.] The whole set of the conserved densities $\{H_r\}$ of the KdV hierarchy can now be *defined* by the bi-Hamiltonian property

$$(7.23)\quad B^I\left(\frac{\delta H_{r+1}}{\delta u}\right) = B^{II}\left(\frac{\delta H_r}{\delta u}\right), \quad r \in \mathbf{Z}_+,$$

with the initial data H_0 satisfying

$$(7.24)\quad B^I\left(\frac{\delta H_0}{\delta u}\right) = 0.$$

Obviously, the relations (7.23), (7.24) by themselves imply that all H_r's are in involution with respect to each one of the Hamiltonian structures B^I and B^{II}. It follows that all the KdV equations mutually commute. Thus, we arrive at the desired re-interpretation of the KdV hierarchy: given two Hamiltonian matrices B^I and B^{II}, and $H_0 \in \text{Ker } B^I$, construct an infinite series $\{H_r\}$ satisfying the bi-Hamiltonian property (7.23), (7.24). This is precisely the point of view we are now going to adopt, with additional proviso that both B^I and B^{II} correspond to a pair of 2-cocycles on an appropriate differential

Lie algebra or superalgebra. Needless to say, absent root space decompositions, highest weight modules and other basic instruments from representation theory of affine Lie algebras, and even such everyday tools as Lax and zero-curvature representations, we shall need every trick we can devise to analyse the main problem which we are now facing, that is, an *existence of an infinite sequence* $\{H_r\}$, given H_0, B^I, and B^{II}.

To proceed further, we now describe the first class of Lie algebras with which the Hamiltonian structures B^I and B^{II} will be associated, for the first case of superintegrable systems attached to a metrizable Lie algebra $\mathcal{G}$. Let $\mathcal{G}$ be a metrizable Lie algebra and a free $\mathcal{A}_0$-module over a commutative ring $\mathcal{A}_0$, $\mathcal{A}_0 \supset \mathbf{Q}$, and let $(\ , \ )$ be the invariant form on $\mathcal{G}$. (All Lie algebras and vector spaces over $\mathcal{A}_0$ are considered as free $\mathcal{A}_0$-modules in this Chapter.) We assume that $(\ , \ )$ and $\mathcal{A}_0$ are such that there exists an orthonormal basis $(e_1, ..., e_n)$ in $\mathcal{G}$. We fix this basis once and for all. Let $K = K_0 + K_1$ be a commutative superalgebra (see §2), and let $\partial \colon K \to K$ be an even derivation. Set $K_c = \operatorname{Ker} \partial\big|_K, K_c = (K_c)_0 + (K_c)_1$, and assume that $(K_c)_0 \supset \mathcal{A}_0$ (this is the same $\mathcal{A}_0$, which, if desired, may be taken to be $\mathbf{Q}$). Let N_0, N_1 be two nonnegative integers with $N_0 + N_1 > 0$. Recall that (see §2) the K-module $K^{N_0|N_1}$ of column-vectors of length $N_0 + N_1$ has the following $\mathbf{Z}_2$-grading:

$$(7.25) \quad p(X) = p(X_i) + p(i)\,, \quad X_i \in K,$$

$$(7.26) \quad p(i) = 0 \in \mathbf{Z}_2, 1 \leq i \leq N_0; \quad p(i) = 1 \in \mathbf{Z}_2, N_0 < i \leq N_0 + N_1.$$

In particular, even vectors in $K^{N_0|N_1}$ are elements of $K_0^{N_0} \oplus K_1^{N_1}$, and $K^{1|0} = K$. A (stable) Lie superalgebra structure $\mathcal{H}$ on $K^{N_0|N_1}$ is an even map

$[\ ,\ ]\colon K^{N_0|N_1} \times K^{N_0|N_1} \to K^{N_0|N_1}$, of the form (see Definition 3.172)

$$(7.27)\quad [X,Y]_{\mathrm{k}} = \sum (-1)^{p(i)p(X)} c^{\mathrm{k}}_{i,\ell|j,m} X_j^{(m)} Y_i^{(\ell)}, \quad c^{\mathrm{k}}_{\dots} \in K,$$

(where $(\cdot)^{(m)} = \partial^m(\cdot)$), satisfying the following properties:

$$(7.28)\quad [X,Y] = -(-1)^{p(X)p(Y)}[Y,X];$$

$$(7.29)\quad [[X,Y],Z] = [X,[Y,Z]] - (-1)^{p(X)p(Y)}[Y,[X,Z]];$$

$$(7.30)\quad [X,Yc] = [X,Y]c, \quad \forall c \in K_c,$$

where $(Yc)_i = Y_i c$; and

(7.31) The properties (7.27)–(7.30) remain true under arbitrary extension $\widetilde{K} \supset K$.

It follows that a Lie superalgebra $\mathcal{H}$ is uniquely defined by the Lie algebra $\mathcal{H}_0$ consisting of even elements in $\mathcal{H}$: $\mathcal{H}_0 = K_0^{N_0} \oplus K_1^{N_1}$. In what follows we shall always work only with such Lie algebras. (In this and in the next Sections, $N = N_0 + N_1$ is always finite.)

Let $\mathcal{F}$ be an n-dimensional Lie algebra over $\mathcal{A}_0$. Denote by $\mathcal{F}_{\mathrm{aff}} = K_0^{n+1}$ and $\mathcal{F}^s_{\mathrm{aff}} = K_0^{n+1} \oplus K_1^{n+1}$ Lie algebras, over K_0 and K respectively, with the following multiplication

$$\left[\begin{pmatrix} X^1 \\ f^1 \otimes a^1 \end{pmatrix}, \begin{pmatrix} X^2 \\ f^2 \otimes a^2 \end{pmatrix}\right] =$$

$$(7.32)\quad \begin{pmatrix} X^1\partial(X^2) - X^2\partial(X^1) \\ X^1\partial(f^2) \otimes a^2 - X^2\partial(f^1) \otimes a^1 + f^1 f^2 \otimes [a^1, a^2] \end{pmatrix}$$

$$X^i, f^i \in K_0, \quad a^i \in \mathcal{F}, \quad i = 1,2$$

(tensor products over A_0 unless specified otherwise),

$$\left[\begin{pmatrix} X^1 \\ f^1\otimes a^1 \\ \gamma^1\otimes b^1 \\ \alpha^1 \end{pmatrix}, \begin{pmatrix} X^2 \\ f^2\otimes a^2 \\ \gamma^2\otimes b^2 \\ \alpha^2 \end{pmatrix}\right] =$$

$$\begin{pmatrix} X^1\partial(X^2) - X^2\partial(X^1) - 2\alpha^1\alpha^2 \\ X^1\partial(f^2)\otimes a^2 - X^2\partial(f^1)\otimes a^1 + f^1f^2\otimes[a^1,a^2] + \gamma^1\alpha^2\otimes b^1 - \gamma^2\alpha^1\otimes b^2 \\ ((X^1\partial(\gamma^2)+\frac{1}{2}\partial(X^1)\gamma^2)\otimes b^2 + f^1\gamma^2\otimes[a^1,b^2] + \partial(f^1)\alpha^2\otimes a^1) - (1\leftrightarrow 2) \\ (X^1\partial(\alpha^2) - \frac{1}{2}\partial(X^1)\alpha^2) - (1\leftrightarrow 2) \end{pmatrix}$$

(7.33)

$$\gamma^i, \alpha^i \in K_1, \quad b^i \in \mathcal{F}, \quad i = 1, 2,$$

where " $-1 \leftrightarrow 2$"stands for "minus the same expression with the indices 1 and 2 interchanged". Thus, $\mathcal{F}_{\text{aff}}$ is a Lie subalgebra in $\mathcal{F}^s_{\text{aff}}$. Denote $D^s = D^s(K) = \{0\}^s_{\text{aff}}$. Noticing that $D = D(K) = \{0\}_{\text{aff}}$, we see that

(7.34) $\mathcal{F}_{\text{aff}} = D \propto (K_0 \otimes \mathcal{F})$

(7.35) $\mathcal{F}^s_{\text{aff}} = D^s \propto ((K_0 \otimes \mathcal{F} \underset{ad}{\propto} (K_1 \otimes V(\mathcal{F})))$

where $\propto$ denotes the semidirect product, and $V(\mathcal{F})$ denotes the $\mathcal{F}$-module of the adjoint representation of $\mathcal{F}$ but considered as an abelian Lie algebra by itself.

We now describe 2-cocycles on Lie algebras $\mathcal{F}_{\text{aff}}$ and $\mathcal{F}^s_{\text{aff}}$. We start with $\mathcal{F}^s_{\text{aff}}$ and then restrict 2-cocycles from $\mathcal{F}^s_{\text{aff}}$ onto its Lie subalgebra $\mathcal{F}_{\text{aff}}$.

7.36. Lemma. The following is a 2-cocycle on $\mathcal{F}^s_{\text{aff}}$:

(7.37) $\omega_1(1,2) = \partial^3(X^1)X^2 + 4\partial^2(\alpha^1)\alpha^2$.

where we write $\omega(1,2)$ instead of $\omega\begin{pmatrix} X^1 & X^2 \\ \vdots & , \vdots \end{pmatrix}$.

Proof. Obviously, ω_1 is skewsymmetric. Now,

$$\omega_1(3,[1,2]) + \text{ c.p } = \partial^3(X^3)[X^1\partial(X^2) - X^2\partial(X^1) - 2\alpha^1\alpha^2] + \text{c.p.} +$$

$$+4\partial^2(\alpha^3)[X^1\partial(\alpha^2) - \tfrac{1}{2}\partial(X^1)\alpha^2 - X^2\partial(\alpha^1) + \tfrac{1}{2}\partial(X^2)\alpha^1] + \text{ c.p. } \sim$$

$$\sim 2X^2\partial^3(\alpha^3\alpha^1) + \text{c.p.} + 4X^2[\partial^2(\alpha^1)\partial(\alpha^3) + \tfrac{1}{2}\partial(\partial^2(\alpha^1)\alpha^3)] + \text{c.p.} -$$

$$-4X^2[\partial^2(\alpha^3)\partial(\alpha^1) + \tfrac{1}{2}\partial(\partial^2(\alpha^3)\alpha^1)] + \text{ c.p. } =$$

$$= 2X^2[\alpha^{3'''}\alpha^1 + 3\alpha^{3''}\alpha^{1'} + 3\alpha^{3'}\alpha^{1''} + \alpha^3\alpha^{1'''} - 2\alpha^{3'}\alpha^{1''} - \alpha^{3'}\alpha^{1''} - \alpha^3\alpha^{1'''} -$$

$$-2\alpha^{3''}\alpha^{1'} - \alpha^{3'''}\alpha^1 - \alpha^{3''}\alpha^{1'}] + \text{ c.p. } = 0 ,$$

where $(\cdot)' := \partial(\cdot)$ in agreement with the old tradition. ∎

7.38. Lemma. The following is a 2-cocycle on $\mathcal{G}^s_{\text{aff}}$:

$$(7.39)\quad \omega_2(1,2) = \partial(f^1)f^2(a^1,a^2) + \gamma^1\gamma^2(b^1,b^2).$$

Proof. Again, ω_2 is clearly skewsymmetric. Then, $\omega_2(3,[1,2]) + \text{ c.p. } =$

$$(7.40)\quad f^{3'}\{X^1 f^{2'}(a^3,a^2) - X^2 f^{1'}(a^3,a^1)\} + \text{c.p.} + f^{3'}f^1f^2(a^3,[a^1,a^2]) + \text{c.p.} +$$

$$(7.41)\quad f^{3'}\{\gamma^1\alpha^2(a^3,b^1) - \gamma^2\alpha^1(a^3,b^2)\} + \text{c.p.} +$$

$$(7.42)\quad \gamma^3\{(X^1\gamma^{2'} + \frac{1}{2}X^{1'}\gamma^2)(b^3,b^2) - (X^2\gamma^{1'} + \frac{1}{2}X^{2'}\gamma^1)(b^3,b^1) + \text{c.p.}\} +$$

$$(7.43)\quad \gamma^3\{f^1\gamma^2(b^3,[a^1,b^2]) - f^2\gamma^1(b^3,[a^2,b^1])\} + \text{c.p.} +$$

$$(7.44)\quad \gamma^3\{f^{1'}\alpha^2(b^3,a^1) - f^{2'}\alpha^1(b^3,a^2)\} + \text{c.p.}$$

The first cyclic sum in (7.40) equals to

$$f^{1'}X^2f^{3'}(a^1,a^3) + \text{ c.p. } - \left[f^{3'}X^2f^{1'}(a^3,a^1) + \text{ c.p.}\right] = 0.$$

The second cyclic sum in (7.40) is

$$(f^{3'}f^1f^2+f^{1'}f^2f^3+f^{2'}f^3f^1)(a^3,[a^1,a^2])=(f^3f^1f^2)'(a^3,[a^1,a^2])\sim 0.$$

Combining (7.41) and (7.44) we obtain

$$f^{3'}\gamma^1\alpha^2(a^3,b^1)+\text{ c.p. }-[f^{1'}\gamma^3\alpha^2(a^1,b^3)+\text{ c.p.}]+\gamma^3f^{1'}\alpha^2(b^3,a^1)+\text{ c.p. }-$$
$$-[\gamma^1f^{3'}\alpha^2(b^1,a^3)+\text{ c.p.}]=0.$$

Working out (7.42) modulo $Im\partial$ we get

$$X^2[\gamma^1\gamma^{3'}-\frac{1}{2}(\gamma^1\gamma^3)')](b^1,b^3)-X^2[\gamma^3\gamma^{1'}-\frac{1}{2}(\gamma^3\gamma^1)'](b^3,b^1)+\text{ c.p. }=0.$$

Finally, (7.43) becomes

$$\gamma^1f^2\gamma^3(b^1,[a^2,b^3])+\text{ c.p. }-[\gamma^3f^2\gamma^1(b^3,[a^2,b^1])+\text{ c.p.}]=0.$$ ∎

7.45. Corollary. The following are 2-cocycles of $\mathcal{F}_{\text{aff}}$ and $\mathcal{G}_{\text{aff}}$ respectively:

(7.46) $\nu_1(1,2)=\partial^3(X^1)X^2$,

(7.47) $\nu_2(1,2)=\partial(f^1)f^2(a^1,a^2)$.

We are now prepared to derive B^I and B^{II}. We start off by calculating $B=B(\mathcal{G}^s_{\text{aff}})$ (see (3.81)). Let $C=C(\mathcal{G}^s_{\text{aff}})=K[u^{(j)},q_i^{(j)},\sigma_i^{(j)},\varphi^{(j)}], j\in\mathbf{Z}_+$, $1\le i\le n=\dim\mathcal{G}$, be a differential commutative superalgebra, with the $\mathbf{Z}_2$-gradings of generators given by the formula

(7.48) $p(u^{(j)})=p(q_i^{(j)})=0\in\mathbf{Z}_2,\quad p(\sigma_i^{(j)})=p(\varphi^{(j)})=1\in\mathbf{Z}_2\,.$

Then, in analogy with (7.17), we have from (7.33)

$$
\begin{aligned}
B(1)^t 2 \sim & -u[X^1\partial(X^2) - X^2\partial(X^1) - 2\alpha^1\alpha^2] - \\
& -\sum q_i[X^1\partial(f_i^2) - X^2\partial(f_i^1) + f_j^1 f_{\mathrm{k}}^2 t^i_{j\mathrm{k}} + \gamma_i^1\alpha^2 - \gamma_i^2\alpha^1] - \\
& -\sum \sigma_i[X^1\partial(\gamma_i^2) + \frac{1}{2}\partial(X^1)\gamma_i^2 - X^2\partial(\gamma_i^1) - \frac{1}{2}\partial(X^2)\gamma_i^1 + f_j^1\gamma_{\mathrm{k}}^2 t^i_{j\mathrm{k}} - f_j^2\gamma_{\mathrm{k}}^1 t^i_{j\mathrm{k}} + \\
& +\partial(f_i^1)\alpha^2 - \partial(f_i^2)\alpha^1] - \varphi[X^1\partial(\alpha^2) - \frac{1}{2}\partial(X^1)\alpha^2 - X^2\partial(\alpha^1) + \frac{1}{2}\partial(X^2)\alpha^1] \sim
\end{aligned}
$$

$$
\begin{array}{llll}
[(\partial u + u\partial)(X^1) + & q_i\partial(f_i^1) + & (\sigma_i\partial - \frac{1}{2}\partial\sigma_i)(\gamma_i^1) & +(\varphi\partial + \frac{1}{2}\partial\varphi)(\alpha^1)](X^2) + \\
[\partial q_j(X^1) - & t^{\mathrm{k}}_{ij} q_{\mathrm{k}} f_i^1 + & t^{\mathrm{k}}_{ji}\sigma_{\mathrm{k}}\gamma_i^1 & -\partial\sigma_j(\alpha^1)](f_j^2) + \\
[(\partial\sigma_j - \frac{1}{2}\sigma_j\partial)(X^1) + & t^{\mathrm{k}}_{ji}\sigma_{\mathrm{k}} f_i^1 & & -q_j\alpha^1](\gamma_j^2) + \\
[(\partial\varphi + \frac{1}{2}\varphi\partial)(X^1) - & \sigma_i\partial(f_i^1) & -q_i\gamma_i^1 & +2u\alpha^1](\alpha^2),
\end{array}
\tag{7.49}
$$

where $t^i_{j\mathrm{k}}$ are the structure constants of $\mathcal{G}$ in the orthonormal basis $(e_1, ..., e_n)$, and summation is (and will always be) implied over repeated indices (unless specified otherwise). Thus, the matrix $B = B(\mathcal{G}^s_{\mathrm{aff}})$ is

$$
B = \begin{pmatrix}
u\partial + \partial u & q_j\partial & \sigma_j\partial - \frac{1}{2}\partial\sigma_j & \varphi\partial + \frac{1}{2}\partial\varphi \\
\partial q_i & t^{\mathrm{k}}_{ij} q_{\mathrm{k}} & t^{\mathrm{k}}_{ij}\sigma_{\mathrm{k}} & -\partial\sigma_i \\
\partial\sigma_i - \frac{1}{2}\sigma_i\partial & t^{\mathrm{k}}_{ij}\sigma_{\mathrm{k}} & 0 & -q_i \\
\partial\varphi + \frac{1}{2}\varphi\partial & -\sigma_j\partial & -q_j & 2u
\end{pmatrix}
\tag{7.50}
$$

We now compute the matrices associated to 2-cocycles on $\mathcal{G}^s_{\mathrm{aff}}$, via the same rule (7.19). We have, from (7.37):

$$
b_1(1)^t 2 \sim \partial^3(X^1)X^2 + 4\partial^2(\alpha^1)\alpha^2,
$$

so that

(7.51) $b_1 = \text{ diag } (\partial^3, \mathbf{0}, \mathbf{0}, 4\partial^2)$.

Analogously, we find from (7.39)

(7.52) $b_2 = \text{ diag } (0, \mathbf{1}\partial, \mathbf{1}, 0)$

Finally, consider the trivial 2-cocycle "component #1" on $\mathcal{G}^s_{\text{aff}}$, that is

(7.53) $\omega_3(1,2) = X^1\partial(X^2) - X^2\partial(X^1) - 2\alpha^1\alpha^2$.

Then

$$\omega_3(1,2) \sim -2\left[\partial(X^1)X^2 + \alpha^1\alpha^2\right],$$

so that

(7.54) $b_3 = -2 \text{ diag } (\partial, \mathbf{0}, \mathbf{0}, 1)$.

Now set

(7.55) $B^I = -\frac{1}{4}b_3 + \frac{1}{2}b_2 = \frac{1}{2} \text{ diag } (\partial, \partial\mathbf{1}, \mathbf{1}, 1)$;

$$B^{II} = B(\mathcal{G}^s_{\text{aff}}) - \frac{1}{2}b_1 =$$

(7.56)
$$= \begin{pmatrix} u\partial + \partial u - \frac{1}{2}\partial^3 & q_j\partial & \sigma_j\partial - \frac{1}{2}\partial\sigma_j & \varphi\partial + \frac{1}{2}\partial\varphi \\ \partial q_i & t^{\mathrm{k}}_{ij}q_{\mathrm{k}} & t^{\mathrm{k}}_{ij}\sigma_{\mathrm{k}} & -\partial\sigma_i \\ \partial\sigma_i - \frac{1}{2}\sigma_i\partial & t^{\mathrm{k}}_{ij}\sigma_{\mathrm{k}} & 0 & -q_i \\ \partial\varphi + \frac{1}{2}\varphi\partial & -\sigma_j\partial & -q_j & 2(u - \partial^2) \end{pmatrix}$$

Set, in analogy with (7.2i),

(7.57) $H_0 = u.$

Then

$B^I\delta(H_0) = \mathbf{0}$, and

$$B^{II}\delta(H_0) = \partial(u, \underline{q}^t, \underline{\sigma}^t, \varphi)^t = B^I(2u, 2\underline{q}^t, 2\underline{\sigma}'^t, 2\varphi')^t = B^I\delta(H_1),$$

where

$\delta(H)$ is the column-vector of variational derivatives of $H \in C$:

(7.58) $$\frac{\delta H}{\delta p} = \sum(-\partial)^j\left(\frac{\delta H}{\delta p^{(j)}}\right),$$

(7.59) $$H_1 = u^2 + (\underline{q}, \underline{q}) + (\underline{\sigma}, \underline{\sigma}') + \varphi\varphi',$$

and $(\underline{p}, \underline{\tilde{p}}) := \sum p_i \tilde{p}_i$. Hence, computing $B^{II}\delta(H_1)$ we arrive at the following generalization of the KdV equation:

$$u_t = (u\partial + \partial u - \frac{1}{2}\partial^3)(2u) + q_j\partial(2q_j) + (\sigma_j\partial - \frac{1}{2}\partial\sigma_j)(2\sigma_j')+$$
$$+(\varphi\partial + \frac{1}{2}\partial\varphi)(2\varphi') = 6uu_x - u_{xxx} + \partial[(\underline{q}, \underline{q}) + (\underline{\sigma}, \underline{\sigma}') + 3\varphi\varphi'],$$
$$q_{i,t} = \partial q_i(2u) + 2t_{ij}^{\mathrm{k}} q_{\mathrm{k}} q_j + 2t_{ij}^{\mathrm{k}}\sigma_{\mathrm{k}}\sigma_j' - 2\partial\sigma_i(\varphi') =$$
$$= \partial(2q_i u - [\underline{\sigma}, \underline{\sigma}]_i - 2\sigma_i\varphi'),$$

(7.60)

$$\sigma_{i,t} = (\partial\sigma_i - \frac{1}{2}\sigma_i\partial)(2u) + 2t_{ij}^{\mathrm{k}}\sigma_{\mathrm{k}} q_j - q_i 2\varphi' =$$
$$= 2u\sigma_i' + \sigma_i u' + 2[\underline{q}, \underline{\sigma}]_i - 2\varphi' q_i,$$
$$\varphi_t = (\partial\varphi + \frac{1}{2}\varphi\partial)(2u) - \sigma_j\partial(2q_j) - q_j 2\sigma_j' + 2(u - \partial^2)(2\varphi') =$$
$$= 6u\varphi' + 3u'\varphi - 4\varphi''' - 2(\underline{q}, \underline{\sigma})',$$

where $[\underline{\sigma},\underline{\sigma}]_i = t^i_{jk}\sigma_j\sigma_k = t^k_{ij}\sigma_j\sigma_k$, since the structure constants t^k_{ij} are cyclic-symmetric in an orthonormal basis. In vector notation, the system (7.60) takes the form

$$(7.61)\quad \begin{cases} u_t = \partial[3u^2 - u_{xx} + (\underline{q},\underline{q}) + (\underline{\sigma},\underline{\sigma}') + 3\varphi\varphi'], \\ \underline{q}_t = \partial\{2u\underline{q} - [\underline{\sigma},\underline{\sigma}] + 2\varphi'\underline{\sigma}\}, \\ \underline{\sigma}_t = 2u\underline{\sigma}' + u'\underline{\sigma} + 2[\underline{q},\underline{\sigma}] - 2\varphi'\underline{q}, \\ \varphi_t = 6u\varphi' + 3u'\varphi - 4\varphi''' - 2(\underline{q},\underline{\sigma})'. \end{cases}$$

Few remarks are in order. First, for $\underline{q} = \underline{\sigma} = \mathbf{0}$, $\varphi = 0$, our $\mathcal{G}$-sKdV system (7.61) becomes just the KdV equation (7.1) itself. Thus, we have constructed a true generalization of the KdV equation for every metrizable Lie algebra $\mathcal{G}$. It remains to show that this generalization is meaningful, i.e., that the bi-Hamiltonian definition

$$(7.62)\quad B^I\delta(H_{r+1}) = B^{II}\delta(H_r),$$

with $H_0 = u$, can be iterated for all $r \in \mathbf{Z}_+$ to produce an infinity of conserved densities of the $\mathcal{G}$-sKdV equation (7.61).The rest of this Chapter, §§8,9, is devoted mainly to this task. At the moment let us notice that (7.62) is satisfied not only for $r = 0$ but also for $r = 1$ if we define

$$(7.63)\quad \begin{aligned} H_2 = 2u^3 + u_x^2 + 2u(\underline{q},\underline{q}) + 2u(\underline{\sigma},\underline{\sigma}') + 6u\,\varphi\varphi' + \\ + 2(\underline{q}, 2\varphi'\underline{\sigma} - [\underline{\sigma},\underline{\sigma}]) - 4\varphi\varphi'''. \end{aligned}$$

Secondly, letting in (7.61) $\underline{\sigma}$ and $\underline{\varphi}$ vanish, results in the purely even system

$$(7.64)\quad \begin{cases} u_t = \partial[3u^2 - u_{xx} + (\underline{q},\underline{q})], \\ \underline{q}_t = \partial(2u\underline{q}), \end{cases}$$

which inherits from $\mathcal{G}$ *only its dimension.* This is the reason why the $\mathbf{Z}_2$-graded detour is important. (An alternative is given by (7.130) below.) Note that the system (7.64) is known ([Ku 15]) to satisfy (7.62) for all $r \in \mathbf{Z}_+$. Thirdly, for the case $\mathcal{G} = \{0\}$, the system (7.61) becomes

$$(7.65)\quad \begin{cases} u_t = \partial(3u^2 - u_{xx} + 3\varphi\varphi_x)\ , \\ \varphi_t = 6u\varphi_x + 3u_x\varphi - 4\varphi_{xxx}. \end{cases}$$

The system (7.65) is also known ([Ku 5]) to have an infinity of conserved densities. However, the proof of this fact in [Ku 5], which makes use of a deformation of the Miura map which the system (7.65) possesses, is not applicable in the general case (7.61) when $\mathcal{G} \neq \{0\}$ since in this case there exists no Miura map. (In the physical language, the $\mathcal{G}$-sKdV system (7.61) is not Galilean invariant unless both $\underline{q}$ and $\underline{\sigma}$ vanish.)

The task of solving the equation (7.62) is typically divided into three general steps. First, consider column-vectors

$$(7.66)\quad X_r \in C_0^{n+1} \oplus C_1^{n+1}, \qquad n = \dim \mathcal{G}, \qquad r \in \mathbf{Z}_+,$$

where $C = C(\mathcal{G}^s_{\text{aff}}) = K[u^{(j)}, q_i^{(j)} \sigma_i^{(j)}, \varphi^{(j)}]$; thus, $X_r \in (C^{n+1|n+1})_0$. Set

$$(7.67)\quad (X_0)_j = \delta_j^1, \quad 1 \leq j \leq 2(n+1),$$

so that $X_0 = \delta(H_0)$, and consider the set of equations on X_r's:

$$(7.68)\quad B^I(X_{r+1}) = B^{II}(X_r),$$

for all $r \in \mathbf{Z}_+$. We want to show that: 2) we can solve (7.68) step by step, starting with the initial data (7.67); 3) and moreover, that the resulting vectors

$\{X_r\}$ are vectors of functional derivatives of some elements $\{H_r\}$ from C. (The latter property, by Theorems 2.65 and 2.67, is equivalent to the commutative Fréchet derivative $D(X_r)$ being supersymmetric for all $r \in \mathbf{Z}_+$.)

Let us see where the difficulties lie. Denote

$$(7.69)\quad X_r = \begin{pmatrix} x_r \\ \underline{a}_r \\ \underline{\psi}_r \\ \chi_r \end{pmatrix}, \quad x_r \in C_0, \underline{a}_r \in C_0^n, \underline{\psi}_r \in C_1^n, \chi_r \in C_1 .$$

Then (7.68) reads, with the help of (7.55) and (7.56):

$$\frac{1}{2} x'_{r+1} = (u\partial + \partial u - \frac{1}{2}\partial^3)(x_r) + (\underline{q}, \underline{a}'_r) + (\underline{\sigma}, \underline{\psi}_r{}') - \frac{1}{2}(\underline{\sigma}, \underline{\psi}_r)' +$$

$$(7.70a)\quad + \varphi\chi'_r + \frac{1}{2}(\varphi\chi_r)' \sim$$

$$(7.70b)\quad \sim u x'_r + (\underline{q}, \underline{a}'_r) + (\underline{\sigma}, \underline{\psi}_r{}') + \varphi\chi'_r .$$

$$(7.71a)\quad \frac{1}{2}\underline{a}'_{r+1} = (\underline{q} x_r - \underline{\sigma}\chi_r)' - [\underline{q}, \underline{a}_r] - [\underline{\sigma}, \underline{\psi}_r] \sim$$

$$(7.71b)\quad \sim -[\underline{q}, \underline{a}_r] - [\underline{\sigma}, \underline{\psi}_r].$$

$$(7.72)\quad \frac{1}{2}\underline{\psi}_{r+1} = (\partial\underline{\sigma} - \frac{1}{2}\underline{\sigma}\partial)(x_r) - [\underline{\sigma}, \underline{a}_r] - \underline{q}\chi_r .$$

$$(7.73)\quad \frac{1}{2}\chi_{r+1} = (\partial\varphi + \frac{1}{2}\varphi\partial)(x_r) - (\underline{\sigma}, \underline{a}'_r) - (\underline{q}, \underline{\psi}_r) + 2(u - \partial^2)(\chi_r),$$

where, as agreed before, $(p, \tilde{p}) := \sum p_i \tilde{p}_i$, for $\mathcal{G} \otimes C$ valued vectors $p, \tilde{p}$.

Thus, (7.72) and (7.73) present no problems, defining $\underline{\psi}_{r+1}$ and χ_{r+1}. As far as (7.70) is concerned, we need the R.H.S. of it to be in $Im\partial$. To show that this is true, and for other reasons which will emerge gradually, we shall set up the induction procedure. The induction assumptions will also be tightened up gradually. The first induction assumption is minimal:

(7.74) The vectors $X_0, \dots, X_r$ from (7.68) are vectors of functional derivatives of some elements $H_0, \dots, H_r$ in C_c.

(Here and in what follows, we denote by C_c the commutative subalgebra $K_c[q_i^{(g|\nu)}]$ of the commutative algebra $K[q_i^{(g|\nu)}]$, see §2.)

7.75. Lemma. In the notation of §2, let $C = K[q_i^{(g|\nu)}]$, $H \in C_c$, and $1 \leq s \leq m$. Then

$$(7.76) \quad \sum_i q_i \partial_s \left(\frac{\delta H}{\delta q_i} \right) \sim 0 .$$

Proof. We have,

$$-\sum q_i \partial_s \left(\frac{\delta H}{\delta q_i} \right) \sim \sum \partial_s(q_i) \frac{\delta H}{\delta q_i} = \sum \partial_s(q_i)(-\partial)^\nu \hat{g}^{-1} \left(\frac{\partial H}{\partial q_i^{(g|\nu)}} \right) \sim$$

$$\sim \sum \hat{g} \partial^\nu \partial_s(q_i) \frac{\partial H}{\partial q_i^{(g|\nu)}} = \sum \partial_s \left(q_i^{(g|\nu)} \right) \frac{\partial H}{\partial q_i^{(g|\nu)}} [\text{ since } H \in C_c] = \partial_s(H) \sim 0. \blacksquare$$

Applying Lemma 7.75 to the R.H.S. of (7.70) and using the induction assumption (7.74), we find that the expression (7.70b), being of the form (7.76), is indeed trivial and, therefore, we can solve (7.70) in favour of x_{r+1}. Let us turn to the only remaining entry in the system (7.68), namely (7.71). To solve for $\underline{a}_{r+1}$, we need the following relation to hold

$$(7.77) \quad [\underline{q}, \frac{\delta H_r}{\delta \underline{q}}] + \left[\underline{\sigma}, \frac{\delta H_r}{\delta \underline{\sigma}} \right] \sim 0 .$$

It is obvious, that (7.77) is not true for *all* $H \in C_c$. Thus we have to significantly fortify our inductive assumptions on the nature of $H_0, \dots, H_r$. A clue can be gotten from an attentive look at (7.59) and (7.63).

Denote $\widetilde{C}_c = K_c[u^{(j)}, \varphi^{(j)}]$, and denote by $\widehat{\mathcal{G}}$ the additive group generated by multiple commutators in $\mathcal{G}$ formed from vectors $\underline{q}^{(j)}$ and $\underline{\sigma}^{(j)}$. Denote by $\mathcal{M}_{\widehat{\mathcal{G}}}$ the ring (and $\mathcal{A}_0$-module) generated (over $\mathcal{A}_0$) by the expressions $\{(y, z) | y, z \in \widehat{\mathcal{G}}\}$ where (,) is the invariant form on $\mathcal{G}$ extended to $\widehat{\mathcal{G}}$. Denote $\mathcal{R} = \widetilde{C}_c \underset{\mathcal{A}_0}{\otimes} \mathcal{M}_{\widehat{\mathcal{G}}}$.

Our inductive assumption on the nature of H_r's is this:

(7.78) $H_0, ..., H_r$ belong to $\mathcal{R}$.

From (7.59) and (7.63) we see that this assumption is true for $r = 0, 1, 2$. We shall prove in §8 that if $H_r \in \mathcal{R}$ then (7.77) is satisfied. Thus, X_{r+1} is constructed. From the general theory of bi-superHamiltonian systems developed in §9, it will follow that $D(X_{r+1})$ is supersymmetric, so that we can find an $H_{r+1} \in C$ such that $X_{r+1} = \delta(H_{r+1})$. However, this relation defines H_{r+1} only modulo Ker $\delta(= Im\partial + K$ by (2.75)), and to sustain the inductive assumption (7.78) we need our H_{r+1} to belong to $\mathcal{R}$. To get around this problem we pick a very specific H_{r+1}, namely the one provided by the formula (2.69):

$$(7.79)\quad H_{r+1} = \int_0^1 dt\, t^{-1} A_t \left\{ux_{r+1} + (\underline{q}, \underline{a}_{r+1}) + (\underline{\sigma}, \underline{\psi_{r+1}}) + \varphi\chi_{r+1}\right\}.$$

From (7.72) and (7.73) we see that the last two terms in (7.79), $(\underline{\sigma}, \underline{\psi_{r+1}}) + \varphi\chi_{r+1}$ belong to $\mathcal{R}$; from (7.70a) and (7.71a) we also see that the total derivatives expressions in $\frac{1}{2}x'_{r+1}$: $(\partial u - \frac{1}{2}\partial^3)(x_r) - \frac{1}{2}\partial(\underline{\sigma}, \underline{\psi_r}) + \frac{1}{2}\partial(\varphi\chi_r)$, and in $\frac{1}{2}\underline{a}'_{r+1}$: $\partial(\underline{q}x_r - \underline{\sigma}\chi_r)$, contribute in (7.79) elements from $\mathcal{R}$, - both of these statements are implied by the obvious property $\dfrac{\delta H}{\delta u}, \dfrac{\delta H}{\delta \varphi} \in \mathcal{R}$ whenever $H \in \mathcal{R}$, and by the following (almost obvious) fact, proven in §8:

(7.80) If $H \in \mathcal{R}$ then $\dfrac{\partial H}{\partial \underline{q}^{(j)}}, \dfrac{\partial H}{\partial \underline{\sigma}^{(j)}}$ belong to $\mathcal{R} \underset{\mathcal{A}_0}{\otimes} \widehat{\mathcal{G}}$.

To proceed further, we need to know the structure of terms of which (7.70b) and (7.71b) are ∂-images of. We start with (7.70b). From the Proof of Lemma 7.75 we see, using 7.80, that (7.70b) belongs to $\partial(\mathcal{R})$. Hence, $x_{r+1} \in \mathcal{R}$. Therefore, the first summand in (7.79), ux_{r+1}, contributes to H_{r+1} an element from $\mathcal{R}$. Next, (7.71b). For the L.H.S. of (7.77) we have

$$[\underline{p}, \frac{\delta H}{\delta \underline{p}}] = \left[\underline{p}, \sum (-\partial)^j \left(\frac{\partial H}{\partial \underline{p}^{(j)}}\right)\right] \text{ [by (7.80)]} \equiv \sum \left[\underline{p}^{(j)}, \frac{\partial H}{\partial \underline{p}^{(j)}}\right] \mod \partial(\mathcal{R} \otimes \widehat{\mathcal{G}})$$

$$\text{[by Corollary 8.29]} \equiv 0 \mod \partial(\mathcal{R} \otimes \widehat{\mathcal{G}}), \quad H \in \mathcal{R}, \underline{p}\colon = \{\underline{q}; \underline{a}\}.$$

Thus, we see that $\underline{a}_{r+1} \in \mathcal{R} \otimes \widehat{\mathcal{G}}$. Hence, the second term in (7.79), $(\underline{q}, \underline{a}_{r+1})$, contributes to H_{r+1} again an element from $\mathcal{R}$. Thus, the inductive assumption (7.78) is advanced one step forward.

In the remainder of this section we describe two more Lie algebras and associated generalizations of the KdV equation.

We start with the observation that if $\rho\colon \mathcal{F} \to End\mathrm{W}$ is a representation then the following formula provides a representation of the Lie algebra $\mathcal{F}_{\mathrm{aff}}$ (7.32) on $K \otimes \mathrm{W}$:

$$\text{(7.82)} \quad \begin{pmatrix} X \\ f \otimes a \end{pmatrix} : g \otimes \mathrm{w} \mapsto (Xg' + \lambda X'g) \otimes \mathrm{w} + fg \otimes \rho(a)(\mathrm{w}),$$

$$g \in K, w \in \mathrm{W}, \lambda \in (K_c)_0.$$

For $\rho = ad$, restricting this representation on $K_1 \otimes \mathcal{F}$, with $\lambda = \frac{1}{2}$, for the case $\mathcal{F} = \mathcal{G}$, we obtain the following semidirect product Lie algebra $\mathcal{G}^1 = \mathcal{G}_{\mathrm{aff}} \propto (K_1 \otimes V(\mathcal{G}))$:

$$\left[\begin{pmatrix} X^1 \\ f^1 \otimes a^1 \\ \gamma^1 \otimes b^1 \end{pmatrix}, \begin{pmatrix} X^2 \\ f^2 \otimes a^2 \\ \gamma^2 \otimes b^2 \end{pmatrix}\right] =$$

(7.83)

$$\begin{pmatrix} X^1\partial(X^2) - X^2\partial(X^1) \\ X^1\partial(f^2)\otimes a^2 - X^2\partial(f^1)\otimes a^1 + f^1 f^2 \otimes [a^1, a^2] \\ ((X^1\partial(\gamma^2) + \frac{1}{2}\partial(X^1)\gamma^2)\otimes b^2 + f^1\gamma^2 \otimes [a^1, b^2]) - (1 \leftrightarrow 2) \end{pmatrix},$$

$X^i, f^i \in K_0, \gamma^i \in K_1, a^i, b^i \in \mathcal{G}$.

Comparing (7.83) with (7.33), we see that $\mathcal{G}^1$ is just a Lie subalgebra in $\mathcal{G}^s_{\text{aff}}$: $\mathcal{G}^1 = \mathcal{G}^s_{\text{aff}}\big|_{\alpha=0}$. Notice that $\mathcal{G}^1$ has the following feature: the 2-cocycle (7.39) on $\mathcal{G}^s_{\text{aff}}$ decouples into a pair of 2-cocycles on $\mathcal{G}^1$, one of them being (7.47) on the subalgebra $\mathcal{G}_{\text{aff}}$ in $\mathcal{G}^1$, and another one being

(7.84) $\omega_4(1,2) = \gamma^1\gamma^2(b^1, b^2)$.

Indeed, $\omega_4([1,2],3)$ + c.p. =

(7.85) $-\gamma^3(X^1\gamma^{2\prime} + \frac{1}{2}X^{1\prime}\gamma^2)(b^2, b^3) + \text{c.p.} +$

$+\gamma^3(X^2\gamma^{1\prime} + \frac{1}{2}X^{2\prime}\gamma^1)(b^1, b^3) + \text{c.p.} -$

(7.86) $-\gamma^3 f^1\gamma^2([a^1, b^2], b^3) + \text{c.p.} + \gamma^3 f^2\gamma^1([a^2, b^1], b^3) + \text{c.p.}$

Working out (7.85), we obtain

$$-\gamma^3(X^1\gamma^{2\prime} + \frac{1}{2}X^1\gamma^2)(b^2, b^3) + \text{c.p.} + \gamma^3(X^2\gamma^{1\prime} + \frac{1}{2}X^{2\prime}\gamma^1)(b^1, b^3) + \text{c.p.} =$$

$$= (b^3, b^1)\left\{-\gamma^1(X^2\gamma^{3\prime} + \frac{1}{2}X^{2\prime}\gamma^3) + \gamma^3(X^2\gamma^{1\prime} + \frac{1}{2}X^{2\prime}\gamma^1)\right\} + \text{c.p.} \sim$$

$$\sim (b^3, b^1)X^2\left\{-\gamma^1\gamma^{3\prime} + \frac{1}{2}(\gamma^1\gamma^3)' + \gamma^3\gamma^{1\prime} - \frac{1}{2}(\gamma^3\gamma^1)'\right\} + \text{c.p.} = 0,$$

while transforming (7.86) we get

$$-f^2\left\{\gamma^1\gamma^3([a^2, b^3], b^1) - \gamma^3\gamma^1([a^2, b^1], b^3)\right\} + \text{c.p. [since } (,) \text{ is invariant]}$$

$$= -f^2\gamma^1\gamma^3\left\{(a^2, [b^3, b^1]) + (a^2, [b^1, b^3])\right\} + \text{c.p.} = 0.$$

∎

(An alternative computation-free Proof would be an observation that $\omega_4 = \omega_2$ (7.39) $-\nu_2$ (7.47).)

Repeating the same steps as we have made deriving (7.55), (7.56), we obtain

$$(7.87) \quad B^I = \frac{1}{2} \operatorname{diag}(\partial, \partial\mathbf{1}, \mathbf{1}),$$

$$(7.88) \quad B^{II} = \begin{pmatrix} u\partial + \partial u - \frac{1}{2}\partial^3 & q_j\partial & \sigma_j\partial - \frac{1}{2}\partial\sigma_j \\ \partial q_i & t_{ij}^{\mathrm{k}} q_{\mathrm{k}} & t_{ij}^{\mathrm{k}} \sigma_{\mathrm{k}} \\ \partial\sigma_i - \frac{1}{2}\sigma_i\partial & t_{ij}^{\mathrm{k}} \sigma_{\mathrm{k}} & 0 \end{pmatrix}.$$

7.89. Remark. Matrices B^I (7.87) and B^{II} (7.88) associated to $\mathcal{G}^1$ (7.83), are the upper left corners of the matrices (7.55) and (7.56), repectively, associated to $\mathcal{G}^s_{\mathrm{aff}} \supset \mathcal{G}^1$. This is not an accident.

7.90. Lemma. Suppose that a Lie algebra $\mathcal{H}_1 = K_0^{N_0} \oplus K_1^{N_1}$ is a Lie subalgebra of $\mathcal{H}_2 = K_0^{N_0+M_0} \oplus K_1^{N_1+M_1}$. (i) If ω_2 is a 2–cocycle on $\mathcal{H}_2$ then the upper left $(N_0 + N_1) \times (N_0 + N_1)$ –corner of the associated matrix b_{ω_2} (see (3.95)) represents the matrix b_{ω_1} of the 2–cocycle ω_1 on $\mathcal{H}_1 : \omega_1 = \omega_2|_{\mathcal{H}_1}$; (ii) If B_2 is the Hamiltonian matrix associated to $\mathcal{H}_2$ then the upper left $(N_0 + N_1) \times (N_0 + N_1)$ –corner of B_2 equals to the Hamiltonian matrix B_1 associated to the Lie algebra $\mathcal{H}_1$.

Proof. (i) is an obviosity. To prove (ii), let us write elements of $\mathcal{H}_2$ in the form $\binom{X}{a}, X \in \mathcal{H}_1, a \in K_0^{M_0} \oplus K_1^{M_1}$. Since $\mathcal{H}_1$ is Lie subalgebra in $\mathcal{H}_2$, the commutator in $\mathcal{H}_2$ has the form

$$(7.91) \quad \left[\begin{pmatrix} X \\ a \end{pmatrix}, \begin{pmatrix} Y \\ b \end{pmatrix}\right] = \begin{pmatrix} [X,Y] + \langle X,b\rangle_1 - \langle Y,a\rangle_1 + \langle a,b\rangle_2 \\ \langle X,b\rangle_3 - \langle Y,a\rangle_3 + \langle a,b\rangle_4 \end{pmatrix},$$

where $<,>$ stands for a bilinear operator of its arguments and $[\,,\,]$ is the commutator in $\mathcal{H}_1$. Denote by $\bar{q} = (q_1, \ldots, q_{N_0+N_1})^t$ and $\bar{p} = (p_1, \ldots, p_{M_0+M_1})^t$ the vectors entering the definition (3.81) of B_2. We have

$$\bar{q}^t([X,Y] + <X,b>_1 - <Y,a>_1 + <a,b>_2) +$$

$$+\bar{p}^t(<X,b>_3 - <Y,a>_3 + <a,b>_4) \sim$$

(7.92)

$$\sim \left[\begin{pmatrix} B_q & -\hat{b}_q^\dagger - \Omega_p^\dagger \\ \hat{b}_q + \Omega_p & \beta_q + \gamma_p \end{pmatrix} \begin{pmatrix} X \\ a \end{pmatrix}\right]^t \begin{pmatrix} Y \\ b \end{pmatrix},$$

where

(7.93) $$\bar{q}^t[X,Y] \sim B_q(X)^t Y,$$

(7.94) $$\bar{q}^t <X,b>_1 \sim \hat{b}_q(X)^t b, \quad \bar{q}^t <a,b>_2 \sim \beta_q(a)^t b,$$

(7.95) $$\bar{p}^t <X,b>_3 \sim \Omega_p(X)^t b, \quad \bar{p}^t <a,b>_4 \sim \gamma_p(a)^t b.$$

In particular, from (7.93) it follows that $B_q = B_1$. ∎

Taking now

(7.96) $$H_0 = u, \; H_1 = u^2 + (\underline{q}, \underline{q}) + (\underline{\sigma}, \underline{\sigma}'),$$

we obtain, as $B^{II}\delta(H_1)$, the following $\mathcal{G}$-sKdV$_1$ system:

(7.97) $$\begin{cases} \dot{u} = \partial[3u^2 - u'' + (\underline{q}, \underline{q}) + (\underline{\sigma}, \underline{\sigma}')], \\ \dot{\underline{q}} = \partial(2u\underline{q} - [\underline{\sigma}, \underline{\sigma}]), \\ \dot{\underline{\sigma}} = 2u\underline{\sigma}' + u'\underline{\sigma} + 2[\underline{q}, \underline{\sigma}]. \end{cases}$$

Also, $B^{II}\delta(H_1) = B^I\delta(H_2)$, where

(7.98) $$H_2 = 2u^3 + u_x^2 + 2u(\underline{q}, \underline{q}) + 2u(\underline{\sigma}, \underline{\sigma}') - 2(\underline{q}, [\underline{\sigma}, \underline{\sigma}]).$$

Thus, H_2 (7.98) equals to $H_2|_{\varphi=0}$ (7.63). Nevertheless, and this is an atypical feature, our $\mathcal{G}$-sKdV$_1$ system (7.97) associated to the Lie subalgebra $\mathcal{G}^1$ of the Lie algebra $\mathcal{G}^s_{\text{aff}}$, is *not* a factor-system $\{\varphi = 0\}$ of the larger $\mathcal{G}$-sKdV system (7.61) associated to the Lie algebra $\mathcal{G}^s_{\text{aff}}$. (Even imposing the condition $(\underline{q}, \underline{\sigma}) = 0$ won't help since this condition, together with the condition $\{\varphi = 0\}$, will be not preserved by the dynamics of (7.61).) It, thus, appears that we put $\varphi = 0$ in the first three equations in (7.61) and, upon realizing that the remaining fourth equation for φ_t becomes a contradiction, discarded it off. In general, such a conduct will result in a worthless remainder; for some reason, which is not entirely clear to me, this has not happened in our particular case. But then again, we can repeat step–by–step the integrability analysis we have performed above for the system (7.61); omitting all the φ– and χ_r–terms during the repetition, we obtain the integrabiltiy of our $\mathcal{G}$-sKdV$_1$ system (7.97).

Like the $\mathcal{G}$-sKdV system (7.61), the $\mathcal{G}$-sKdV$_1$ system (7.97) inherits from $\mathcal{G}$ only its dimension when we let all the odd variables vanish, resulting in the system (7.64). Our third system will be both nonabelian and even (= commutative, in the $\mathbf{Z}_2$-sense).

Let $\{V_k\}$ be a finite collection of finite-dimensional free $\mathcal{A}_0$-modules, let $\rho_k : \mathcal{G} \to End(V_k)$ be representations of our metrizable Lie algebra $\mathcal{G}$, and let $\rho_k^\dagger : \mathcal{G} \to End(V_k^*)$ be the dual representations:

$$(7.99) \quad < a.v_k^*, v_k > = - < v_k^*, a.v_k >, \qquad a \in \mathcal{G},\ v_k^* \in V_k^*, v_k \in V_k,$$

where $a.w$ denotes $\theta(a)(w)$ for a representation $\theta : \mathcal{G} \to End(W)$, and $<, >$ denotes the natural pairing between W^* and W. Let us fix two sets of numbers

$$(7.100) \quad \{\lambda_k, \mu_k | \lambda_k, \mu_k \in K_c,\ \lambda_k + \mu_k = 1,\ \forall k\},$$

where $K = K_0$ (no odd elements are present in the current construction). Taking the direct sum of representations (7.82) of $\mathcal{G}_{\mathrm{aff}}$ on $(\oplus V_{\mathrm{k}}) \oplus (\oplus V_{\mathrm{k}}^*)$, parametrized by $\{\lambda_{\mathrm{k}}\}$ and $\{\mu_{\mathrm{k}}\}$ we arrive at the following semidirect product Lie algebra $\mathcal{G}^2$:

$$\left[\begin{pmatrix} X^1 \\ f^1 \otimes a^1 \\ g_{\mathrm{k}}^1 \otimes v_{\mathrm{k}}^1 \\ h_{\mathrm{k}}^1 \otimes w_{\mathrm{k}}^1 \end{pmatrix}, \begin{pmatrix} X^2 \\ f^2 \otimes a^2 \\ g_{\mathrm{k}}^2 \otimes v_{\mathrm{k}}^2 \\ h_{\mathrm{k}}^2 \otimes w_{\mathrm{k}}^2 \end{pmatrix} \right] =$$

$$= \begin{pmatrix} X^1 X^{2\prime} - X^{1\prime} X^2 \\ f^1 f^2 \otimes [a^1, a^2] + X^1 f^{2\prime} \otimes a^2 - X^2 f^{1\prime} \otimes a^1 \\ \left((X^1 g_{\mathrm{k}}^{2\prime} + \lambda_{\mathrm{k}} X^{1\prime} g_{\mathrm{k}}^2) \otimes v_{\mathrm{k}}^2 + f^1 g_{\mathrm{k}}^2 \otimes a^1 . v_{\mathrm{k}}^2\right) - (1 \leftrightarrow 2) \\ \left((X^1 h_{\mathrm{k}}^{1\prime} + \mu_{\mathrm{k}} X^{1\prime} h_{\mathrm{k}}^2) \otimes w_{\mathrm{k}}^2 + f^1 h_{\mathrm{k}}^2 \otimes a^1 . w_{\mathrm{k}}^2\right) - (1 \leftrightarrow 2) \end{pmatrix},$$

(7.101) $\quad X^i, f_i, g_{\mathrm{k}}^i, h_{\mathrm{k}}^i \in K, \quad a^i \in \mathcal{G}, v_{\mathrm{k}}^i \in V_{\mathrm{k}}, w_{\mathrm{k}}^i \in V_{\mathrm{k}}^*.$

7.102. <u>Remark</u>. The Lie algebra $\mathcal{G}^2$ (7.101) also remains such without the conditions $\{\lambda_{\mathrm{k}} + \mu_{\mathrm{k}} = 1\}$ in (7.100). The reason for these conditions will become clear from the following statement.

7.103. <u>Lemma</u>. The following form ω^{k} is a 2–cocycle on $\mathcal{G}^2$:

(7.104) $\omega^{\mathrm{k}}(1,2) = g_{\mathrm{k}}^1 h_{\mathrm{k}}^2 < w_{\mathrm{k}}^2, v_{\mathrm{k}}^1 > - g_{\mathrm{k}}^2 h_{\mathrm{k}}^2 < w_{\mathrm{k}}^1, v_{\mathrm{k}}^2 >$ (no sum on k).

<u>Proof</u>. Using the following obvious relation

(7.105) $g(Xh' + \lambda X'h) \sim -(Xg' + \mu X'g)h, \ X, g, h \in K, \ \lambda, \mu \in K_c, \ \lambda + \mu = 1,$

we have: $\omega^{k}([1,2],3)$ + c.p. =

$$(7.106)\quad h_k^3(X^1 g_k^{2\prime} + \lambda_k X^{1\prime} g_k^2) < w_k^3, v_k^2 > \ + \text{c.p.} \ -$$

$$(7.107)\quad - h_k^3(X^2 g_k^{1\prime} + \lambda_k X^{2\prime} g_k^1) < w_k^3, v_k^1 > \ + \text{c.p.} \ +$$

$$(7.108)\quad + h_k^3 f^1 g_k^2 < w_k^3, a^1.v_k^2 > \ + \text{c.p.} \ -$$

$$(7.109)\quad - h_k^3 f^2 g_k^1 < w_k^3, a^2.v_k^1 > \ + \text{c.p.} \ -$$

$$(7.110)\quad - g_k^3(X^1 h_k^{2\prime} + \mu_k X^{1\prime} h_k^2) < w_k^2, v_k^3 > \ + \text{c.p.} \ +$$

$$(7.111)\quad + g_k^3(X^2 h_k^{1\prime} + \mu_k X^{2\prime} h_k^1) < w_k^1, v_k^3 > \ + \text{c.p.} \ -$$

$$(7.112)\quad - g_k^3 f^1 h_k^2 < a^1.w_k^2, v_k^3 > \ + \text{c.p.} \ +$$

$$(7.113)\quad + g_k^3 f^2 h_k^1 < a^2.w_k^1, v_k^3 > \ + \text{c.p.}$$

Taking (7.106) and (7.111) together, we obtain

$$\left[h_k^1(X^2 g_k^{3\prime} + \lambda_k X^{2\prime} g_k^3) < w_k^1, v_k^3 > + g_k^3(X^2 h_k^{1\prime} + \mu_k X^{2\prime} h_k^1) < w_k^1, v_k^3 >\right]$$
$$+ \text{c.p. [by (7.105)]} \sim 0.$$

Collecting (7.107) and (7.110), we get, for the same reason

$$\left\{- < w_k^3, v_k^1 > \left[h_k^3(X^2 g_k^{1\prime} + \lambda_k X^{2\prime} g_k^1) + g_k^1(X^2 h_k^{3\prime} + \mu_k X^{2\prime} h_k^3)\right]\right\} \ + \text{c.p.} \ \sim 0.$$

Now, (7.108) and (7.113) join into

$$\left\{h_k^1 f^2 g_k^3 \left[< w_k^1, a^2.v_k^3 > + < a^2.w_k^1, v_k^3 >\right]\right\} \ + \text{c.p. [by (7.99)]} = 0.$$

Finally, (7.109) and (7.112) result into

$$\left\{-h_k^3 f^2 g_k^1 \left[< w_k^3, a^2.v_k^1 > + < a^2.w_k^3, v_k^1 >\right]\right\} \ + \text{c.p. [by (7.99)]} = 0. \quad \blacksquare$$

We now derive B^I and B^{II} associated with $\mathcal{G}^2$. For B^I, we take the 2-cocycle "component #1" plus ν_2 (7.47) plus $\sum_{\mathrm{k}} \omega^{\mathrm{k}}$ (7.104), resulting in

$$(7.114)\qquad B^I = \begin{pmatrix} & u & \underline{q} & \underline{P}^{\ell} & \underline{Q}^{\ell} \\ u & \partial & & & \\ \underline{q} & & \partial\mathbf{1} & & \\ \underline{P}^{\mathrm{k}} & & & & \mathbf{1}\delta^{\ell}_{\mathrm{k}} \\ \underline{Q}^{\mathrm{k}} & & & -\mathbf{1}\delta^{\ell}_{\mathrm{k}} & \end{pmatrix},$$

where we set

$$(7.115)\quad C = C(\mathcal{G}^2) = K\left[u^{(j)}, q_i^{(j)}, P_\alpha^{\mathrm{k}(j)}, Q_\alpha^{\mathrm{k}(j)}\right].$$

Next we compute $B(\mathcal{G}^2)$ in the manner similar to the derivation of (7.49). Denote by ${}_{\mathrm{k}}\tau^{\beta}_{i\alpha}$ the constants defining the representation ρ_{k}:

$$(7.116)\quad [\rho_{\mathrm{k}}(a)(v)]_\beta = {}_{\mathrm{k}}\tau^{\beta}_{i\alpha} a_i v_\alpha, \qquad a \in \mathcal{G},\ v \in V_{\mathrm{k}}.$$

Then, for the dual representation $\rho^{\dagger}_{\mathrm{k}}$, we get

$$(7.117)\quad \left[\rho^{\dagger}_{\mathrm{k}}(a)(w)\right]_\beta = -{}_{\mathrm{k}}\tau^{\alpha}_{i\beta} a_i w_\alpha, \qquad a \in \mathcal{G},\ w \in V^*_{\mathrm{k}},$$

where w_α are the coordinates of w in the basis of V^*_{k} dual to the basis in V_{k} in which the coordinates of v are denoted by v_α in (7.116). Now,

$$B(1)^t 2 \sim -u(X^1 X^{2\prime} - X^{1\prime} X^2) - q_i(X^1 f_i^{2\prime} - X^2 f_i^{1\prime} + t^i_{ab} f^1_a f^2_b) -$$
$$-P^{\mathrm{k}}_\alpha \left[(X^1 p^{2\mathrm{k}\prime}_\alpha + \lambda_{\mathrm{k}} X^{1\prime} p^{2\mathrm{k}}_\alpha) - (X^2 p^{1\mathrm{k}\prime}_\alpha + \lambda_{\mathrm{k}} X^{2\prime} p^{1\mathrm{k}}_\alpha) + {}_{\mathrm{k}}\tau^{\alpha}_{i\beta}(f^1_i p^{2\mathrm{k}}_\beta - f^2_i p^{1\mathrm{k}}_\beta)\right] -$$

$$-Q^{\mathrm{k}}_\alpha\left[(X^1\pi^{2\mathrm{k}\prime}_\alpha+\mu_{\mathrm{k}}X^{1\prime}\pi^{2\mathrm{k}\prime}_\alpha)-(X^2\pi^{1\mathrm{k}\prime}_\alpha+\mu_{\mathrm{k}}X^{2\prime}\pi^{1\mathrm{k}}_\alpha)-{}_{\mathrm{k}}\tau^{\beta}_{i\alpha}(f^1_i\pi^{2\mathrm{k}}_\beta-f^2_i\pi^{1\mathrm{k}}_\beta)\right]\sim$$

$$\begin{aligned}
&[\,(u\partial+\partial u)(X^1)+q_i\partial(f^1_i)\\
&\quad+(P^{\mathrm{k}}_\alpha\partial-\lambda_{\mathrm{k}}\partial P^{\mathrm{k}}_\alpha)(p^{1\mathrm{k}}_\alpha)+(Q^{\mathrm{k}}_\alpha\partial-\mu_{\mathrm{k}}Q^{\mathrm{k}}_\alpha)(\pi^{1\mathrm{k}}_\alpha)\,](X^2)+\\
&[\,\partial q_j(X^1)-t^s_{ij}q_sf^1_i+P^{\mathrm{k}}_\gamma\,{}_{\mathrm{k}}\tau^{\gamma}_{j\alpha}\,p^{1\mathrm{k}}_\alpha-Q^{\mathrm{k}}_\gamma\,{}_{\mathrm{k}}\tau^{\alpha}_{j\gamma}\,\pi^{1\mathrm{k}}_\alpha\,](f^2_j)+\\
&[\,(\partial P^\ell_\beta-\lambda_\ell P^\ell_\beta\partial)(X^1)-{}_\ell\tau^{\gamma}_{i\beta}P^\ell_\gamma f^1_i\,](p^{2\ell}_\beta)+\\
&[\,(\partial Q^\ell_\beta-\mu_\ell Q^\ell_\beta\partial)(X^1)+{}_\ell\tau^{\beta}_{i\gamma}Q^\ell_\gamma f^1_i\,](\pi^{2\ell}_\beta).
\end{aligned}\tag{7.118}$$

Thus, the matrix $B(\mathcal{G}^2)$ is

(7.119)

$$\begin{pmatrix}
u\partial+\partial u & q_i\partial & P^{\mathrm{k}}_\alpha\partial-\lambda_{\mathrm{k}}\partial P^{\mathrm{k}}_\alpha & Q^{\mathrm{k}}_\alpha\partial-\mu_{\mathrm{k}}\partial Q^{\mathrm{k}}_\alpha\\
\partial q_j & -t^s_{ij}q_s & {}_{\mathrm{k}}\tau^{\gamma}_{j\alpha}P^{\mathrm{k}}_\gamma & -{}_{\mathrm{k}}\tau^{\alpha}_{j\gamma}Q^{\mathrm{k}}_\gamma\\
\partial P^\ell_\beta-\lambda_\ell P^\ell_\beta\partial & -{}_\ell\tau^{\gamma}_{i\beta}P^\ell_\gamma & 0 & 0\\
\partial Q^\ell_\beta-\mu_\ell Q^\ell_\beta\partial & {}_\ell\tau^{\beta}_{i\gamma}Q^\ell_\gamma & 0 & 0
\end{pmatrix}\quad(\text{no sum on } \mathrm{k},\ell).$$

Again, we set

$$B^{II}=B(\mathcal{G}^2)-\operatorname{diag}\left(\frac{1}{2}\partial^3,\mathbf{0},\mathbf{0},\mathbf{0}\right).\tag{7.120}$$

Now we can derive the corresponding KdV–type equation. Let

$$H_0=u,\tag{7.121}$$

so that $B^1\delta(H_0)=\mathbf{0}$. Then $B^{II}\delta(H_0)=(u',\underline{q}',\underline{P}^{\ell\prime},\underline{Q}^{\ell\prime})=B^I\delta(H_1)$, where

$$H_1=\frac{u^2}{2}+\frac{1}{2}(\underline{q},\underline{q})+\sum_{\mathrm{k}}<\underline{P}^{\mathrm{k}\prime},\underline{Q}^{\mathrm{k}}>,\tag{7.122}$$

where $<\underline{x}^{\mathrm{k}},\underline{y}^{\mathrm{k}}> = \sum_{\alpha=1}^{\dim V_{\mathrm{k}}} x_\alpha^{\mathrm{k}} y_\alpha^{\mathrm{k}}$. At the moment, we by fiat declare

$$(7.123)\quad \underline{P}^{\mathrm{k}} \in V_{\mathrm{k}}^* \otimes C, \qquad \underline{Q}^{\mathrm{k}} \in V_{\mathrm{k}} \otimes C,$$

thereby allowing the usual meaning for the notation $<,>$ in (7.122). (This is as yet a harmless declaration whose significance will become clear later on.) Computing $B^{II}\delta(H_1)$ we obtain

$$u_t = 3uu' - \frac{1}{2}u''' + (\underline{q},\underline{q}') + (P_\alpha^{\mathrm{k}}\partial - \lambda_{\mathrm{k}}\partial P_{\mathrm{k}}^\alpha)(-Q_\alpha^{\mathrm{k}\prime}) + (Q_\alpha^{\mathrm{k}}\partial - \mu_{\mathrm{k}}\partial Q_\alpha^{\mathrm{k}})(P_\alpha^{\mathrm{k}\prime}) =$$
$$= \partial\left\{\frac{3}{2}u^2 - \frac{1}{2}u'' + \frac{1}{2}(\underline{q},\underline{q}) + (\lambda_{\mathrm{k}}-1) < \underline{P}^{\mathrm{k}},\underline{Q}^{\mathrm{k}\prime} > -(\mu_{\mathrm{k}}-1) < \underline{P}^{\mathrm{k}\prime},\underline{Q}^{\mathrm{k}} >\right\},$$
$$(7.124)$$

$$q_{j,t} = \partial(q_j u) - t_{ij}^s q_s q_i + {}_{\mathrm{k}}\tau_{j\alpha}^\gamma\, P_\gamma^{\mathrm{k}}(-Q_\alpha^{\mathrm{k}\prime}) - {}_{\mathrm{k}}\tau_{j\gamma}^\alpha\, Q_\gamma^{\mathrm{k}}\, P_\alpha^{\mathrm{k}\prime} =$$
$$(7.125)\quad = \partial(q_j u - {}_{\mathrm{k}}\tau_{j\alpha}^\gamma\, Q_\alpha^{\mathrm{k}}\, P_\gamma^{\mathrm{k}}),$$
$$(7.126)\quad P_{\beta,t}^\ell = (\partial P_\beta^\ell - \lambda_\ell P_\beta^\ell \partial)(u) - {}_\ell\tau_{i\beta}^\gamma\, P_\gamma^\ell q_i,$$
$$(7.127)\quad Q_{\beta,t}^\ell = (\partial Q_\beta^\ell - \mu_\ell Q_\beta^\ell \partial)(u) + {}_\ell\tau_{i\gamma}^\beta\, Q_\gamma^\ell q_i.$$

The second terms in the R.H.S.'s of (7.126) and (7.127) can be rewritten with the help of (7.117), (7.116) as $(\underline{q}.\underline{P}^\ell)_\beta$ and $(\underline{q}.\underline{Q}^\ell)_\beta$ respectively, *provided* we accept (7.123). The second term in the R.H.S. of (7.125) can be put into the form $-\underline{Q}^{\mathrm{k}} \bigtriangledown \underline{P}^{\mathrm{k}}$, where the map $V_{\mathrm{k}} \bigtriangledown V_{\mathrm{k}}^* \to \mathcal{G}$ is defined by the rule

$$(7.128)\quad (x \bigtriangledown y, a) = < y, a.x >, \qquad x \in V_{\mathrm{k}},\ y \in V_{\mathrm{k}}^*,\ a \in \mathcal{G}.$$

In coordinates, we have

$$< y, a.x > \text{ [by (7.116)]} = y_\beta\, {}_{\mathrm{k}}\tau_{i\alpha}^\beta\, a_i x_\alpha = ({}_{\mathrm{k}}\tau_{i\alpha}^\beta\, y_\beta x_\alpha) a_i,$$

so that

(7.129) $(x \nabla y)_i = {}_{\mathrm{k}}\tau^{\beta}_{i\alpha}\, x_\alpha\, y_\beta\,.$

All told, we can put out system (7.124) – (7.127) into the vector form

(7.130)
$$\begin{cases} u_t = \partial\left\{\frac{3}{2}u^2 - \frac{1}{2}u'' + \frac{1}{2}(\underline{q},\underline{q}) + \lambda_{\mathrm{k}} < \underline{P}^{\mathrm{k}\prime}, \underline{Q}^{\mathrm{k}} > -\mu_{\mathrm{k}} < \underline{P}^{\mathrm{k}}, \underline{Q}^{\mathrm{k}\prime} >\right\}, \\ \underline{q}_t = \partial(u\underline{q} - \underline{Q}^{\mathrm{k}} \nabla \underline{P}^{\mathrm{k}}), \\ \underline{P}^{\ell}_{,t} = (\partial \underline{P}^{\ell} - \lambda_\ell \underline{P}^{\ell}\partial)(u) + \underline{q}.\underline{P}^{\ell}, \\ \underline{Q}^{\ell}_{,t} = (\partial \underline{Q}^{\ell} - \mu_\ell \underline{Q}^{\ell}\partial)(u) + \underline{q}.\underline{Q}^{\ell}, \end{cases}$$

and call it the $\mathcal{G}_\rho$-KdV system. The second conserved density H_2 is easily found to be

(7.131)
$$H_2 = \frac{1}{2}u^3 + \frac{1}{4}u_x^2 + \frac{1}{2}u(\underline{q},\underline{q}) - \mu_{\mathrm{k}} u < \underline{P}^{\mathrm{k}}, \underline{Q}^{\mathrm{k}\prime} > +\lambda_{\mathrm{k}} u < \underline{P}^{\mathrm{k}\prime}, \underline{Q}^{\mathrm{k}} > + \\ + < \underline{q}.\underline{P}^{\mathrm{k}}, \underline{Q}^{\mathrm{k}} >,$$

so that $B^I\delta(H_2) = B^{II}\delta(H_1)$.

Let us turn to the integrability problem for the $\mathcal{G}_\rho$-KdV system (7.130). Denote

(7.132)
$$X_r = \begin{pmatrix} x_r \\ \underline{a}_r \\ \underline{y}^{\mathrm{k}}_r \\ \underline{z}^{\mathrm{k}}_r \end{pmatrix}, \qquad \begin{matrix} x_r \in C,\ \underline{a}_r \in \mathcal{G}\otimes C,\ \underline{y}^{\mathrm{k}}_r \in V_{\mathrm{k}} \otimes C, \\ \underline{z}^{\mathrm{k}}_r \in V^*_{\mathrm{k}} \otimes C,\ r \in \mathbf{Z}_+, \end{matrix}$$

with the intial condition $X_0 = \delta(H_0)$:

(7.133) $x_0 = 1, \quad \underline{a}_0 = \mathbf{0}, \quad \underline{y}^{\mathrm{k}}_0 = \mathbf{0}, \quad \underline{z}^{\mathrm{k}}_0 = \mathbf{0}.$

We wish to solve the sequence of equations

(7.134) $B^{I}(X_{r+1}) = B^{II}(X_r)$

for all $r \in \mathbf{Z}_+$, and to have for all r the Fréchet derivative $D(X_r)$ to be symmetric, thus guaranteeing an existence of $\{H_r \in C\}$ such that $X_r = \delta(H_r)$. The symmetric property of $D(X_r)$ follows from the general theory of bi-super-Hamiltonian system developed in §9. Let us, then, concentrate on the question of solvability of the equation (7.134). Using formulae (7.114), (7.119), (7.120), (7.123), and (7.129), we rewrite (7.134) in long hand as

$$x'_{r+1} = (u\partial + \partial u - \frac{1}{2}\partial^3)(x_r) + (\underline{q}, \underline{a}'_r) + < \underline{P}^{\mathrm{k}}, \underline{y}_r^{\mathrm{k}\prime} > -\lambda_{\mathrm{k}} < \underline{P}^{\mathrm{k}}, \underline{y}_r^{\mathrm{k}} >' +$$

(7.135a) $+ < \underline{z}_r^{\mathrm{k}\prime}, \underline{Q}^{\mathrm{k}} > -\mu_{\mathrm{k}} < \underline{z}_r^{\mathrm{k}}, \underline{Q}^{\mathrm{k}} >' \sim$

(7.135b) $\sim ux'_r + (\underline{q}, \underline{a}'_r) + < \underline{P}^{\mathrm{k}}, \underline{y}_r^{\mathrm{k}\prime} > + < \underline{z}_r^{\mathrm{k}\prime}, \underline{Q}^{\mathrm{k}} >,$

(7.136a) $\underline{a}'_{r+1} = (x_r \underline{q})' - [\underline{q}, \underline{a}_r] + \underline{y}_r^{\mathrm{k}} \bigtriangledown \underline{P}^{\mathrm{k}} - \underline{Q}^{\mathrm{k}} \bigtriangledown \underline{z}_r^{\mathrm{k}} \sim$

(7.136b) $\sim -[\underline{q}, \underline{a}_r] + \underline{y}_r^{\mathrm{k}} \bigtriangledown \underline{P}^{\mathrm{k}} - \underline{Q}^{\mathrm{k}} \bigtriangledown \underline{z}_r^{\mathrm{k}},$

(7.137) $\underline{z}_{r+1}^{\mathrm{k}} = (\partial \underline{P}^{\mathrm{k}} - \lambda_{\mathrm{k}} \underline{P}^{\mathrm{k}} \partial)(x_r) + \underline{a}_r . \underline{P}^{\mathrm{k}},$

(7.138) $-\underline{y}_{r+1}^{\mathrm{k}} = (\partial \underline{Q}^{\mathrm{k}} - \mu_{\mathrm{k}} \underline{Q}^{\mathrm{k}} \partial)(x_r) + \underline{a}_r . \underline{Q}^{\mathrm{k}}.$

We see that (7.137) and (7.138) present no problems, and that (7.135b) is solvable by Lemma 7.75 provided we make the first inductive assumption

(7.139) $X_0, \ldots, X_r$ are vectors of functional derivatives of some elements $H_0, \ldots, H_r$ in $C_c = K_c[u^{(j)}, q_i^{(j)}, P_\alpha^{\mathrm{k}(j)}, Q_\alpha^{\mathrm{k}(j)}]$.

The only remaining equation, (7.136), requires for its solvability that

(7.140) $$-\left[\underline{q}, \frac{\delta H}{\delta \underline{q}}\right] + \frac{\delta H}{\delta \underline{P}^{\mathrm{k}}} \bigtriangledown \underline{P}^{\mathrm{k}} - \underline{Q}^{\mathrm{k}} \bigtriangledown \frac{\delta H}{\delta \underline{Q}^{\mathrm{k}}} \sim \mathbf{0}$$

for $H = H_r$, a generalization of (7.77). Obviously, (7.140) is *not* satisfied for arbitrary $H \in C$. Thus, we need to drastically tighten up our assumptions on the nature of H_r's.

Denote $\widetilde{C}_c = K_c[u^{(j)}]$, and denote by $\widehat{\mathcal{G}}^1$ the additive group generated by multiple commutators in $\mathcal{G}$ formed from the vectors $\underline{q}^{(j)}$. Denote by $\mathcal{M}^1_{\widehat{\mathcal{G}},p}$ the ring generated by the expressions $\{(y,z)|y,z \in \widehat{\mathcal{G}}^1\}$, $\{< \underline{P}^{\mathrm{k}(j_1)}, \underline{Q}^{\mathrm{k}(j_2)} >$ (no sum on k)$\}$, and $\{< y.\underline{P}^{\mathrm{k}(j_1)}, \underline{Q}^{\mathrm{k}(j_2)} >$ (no sum on k) $|y \in \widehat{\mathcal{G}}^1\}$. Denote $\mathcal{R}^1 = \widetilde{C}_c \underset{\mathcal{A}_0}{\otimes} \mathcal{M}^1_{\widehat{\mathcal{G}},p}$. Our tentative inductive assumption about H_r's is this:

(7.141) $H_0, \ldots, H_r$ belong to $\mathcal{R}^1$.

From (7.121), (7.122), and (7.131) we see that the assumption (7.141) is true for $r = 0, 1, 2$. In §8 we shall prove, in particular, that if $H \in \mathcal{R}^1$ then (7.140) holds true. Therefore, we can find $\underline{a}_{r+1}$ from (7.136) and, thus, complete the induction step (7.134). To complete the induction step (7.141), it remains to show that the thus obtained vector X_{r+1} is of the form $\delta(H_{r+1})$ for some $H_{r+1} \in \mathcal{R}^1$. As above for the $\mathcal{G}$-sKdV system (7.61), we use formula (2.69) to reconstruct H_{r+1} from X_{r+1}:

(7.142)
$$H_{r+1} = \int_0^1 dt\, t^{-1} A_t \left\{ u x_{r+1} + (\underline{q}, \underline{a}_{r+1}) + < \underline{P}^{\mathrm{k}}, \underline{y}^{\mathrm{k}}_{r+1} > + < \underline{z}^{\mathrm{k}}_{r+1}, \underline{Q}^{\mathrm{k}} > \right\}.$$

Since obviously $\dfrac{\delta H}{\delta u} \in \mathcal{R}^1$ for $H \in \mathcal{R}^1$, we find that $x_r \in \mathcal{R}^1$. Hence, from (7.138) and (7.137) we obtain, for the last two summands in (7.142):

$$< \underline{P}^{k}, \underline{y}^{k}_{r+1} > + < \underline{z}^{k}_{r+1}, \underline{Q}^{k} > \equiv - < \underline{P}^{k}, \underline{a}_{r}.\underline{Q}^{k} > + < \underline{a}_{r}.\underline{P}^{k}, \underline{Q}^{k} >$$

$(\text{mod } \mathcal{R}^{1})$[by (7.99)]

(7.143) $\equiv 2 < a_{r}.\underline{P}^{k}, \underline{Q}^{k} > (\text{ mod} \mathcal{R}^{1})$.

On the other hand, from (7.131) we can compute $\underline{a}_2$, using (7.128):

(7.144) $\underline{a}_2 = u\underline{q} - \underline{Q}^{k} \bigtriangledown \underline{P}^{k}$.

Hence, for $r = 2$ (7.143) becomes $2 < \underline{P}^{\ell}, (\underline{Q}^{k} \bigtriangledown \underline{P}^{k}).\underline{Q}^{\ell} >$ (mod $\mathcal{R}^1$), and this term can not be balanced out by the first two summands in (7.142). The moral is that the inductive assumption (7.141) is too restrictive and can not be sustained. To formulate a more precise assumption we need the following fact.

7.145. Lemma. If $a \in \mathcal{G} \otimes C$, $v \in V_{k} \otimes C$, $w \in V_{k}^{*} \otimes C$, then

(7.146) $[a, v \bigtriangledown w] = (a.v) \bigtriangledown w + v \bigtriangledown (a.w)$.

Proof. For any $d \in \mathcal{G} \otimes C$, we have
$(d, [a, v\bigtriangledown w])$ [since (,) is invariant] $= ([d, a], v\bigtriangledown w)$ [by (7.128)] $=< w, [d, a].v >$ $=< w, d.(a.v) - a.(d.v) >$ [by (7.99)] $=< w, d.(a.v) > + < a.w, d.v >$ [by (7.128)] $= (d, (a.v) \bigtriangledown w) + (d, v \bigtriangledown (a.w)) = (d, (a.v) \bigtriangledown w + v \bigtriangledown (a.w))$.

Since d is arbitrary and (,) is nondegenerate, (7.146) follows. ∎

7.147. Remark. In the context of metrizable Lie algebras, it is often handy to use the method employed in the above Proof: to check an identity in $\mathcal{G} \otimes C$, we take the scalar product of this identity with an arbitrary element $d \in \mathcal{G} \otimes C$, and then check the resulting equality.

Let $\{\mathcal{H}, \mathcal{A}^{k+}, \mathcal{A}^{k-}\}$ be the smallest set satisfying the following properties:

(7.148a) $\mathcal{H} \subset \mathcal{G} \otimes C$ is a Lie algebra over $\mathcal{A}_0$;

(7.148b) $\mathcal{A}^{k+} \subset V_k \otimes C$ and $\mathcal{A}^{k-} \subset V_k^* \otimes C$ are $\mathcal{A}_0-$ and $\mathcal{H}-$modules;

(7.148c) $\mathcal{H} \supset \{\underline{q}^{(j)}\}$; $\mathcal{A}^{k+} \supset \{\underline{Q}^{k(j)}\}$; $\mathcal{A}^{k-} \supset \{\underline{P}^{k(j)}\}$;

(7.148d) $\mathcal{A}^{k+} \bigtriangledown \mathcal{A}^{k-} \subset \mathcal{H}$ (no sum on k).

Denote $\widetilde{\mathcal{H}} = \mathcal{H} \underset{\mathcal{A}_0}{\otimes} \widetilde{C}_c$, and analogously for $\widetilde{\mathcal{A}}^{k+}$ and $\widetilde{\mathcal{A}}^{k-}$. (Recall that $\widetilde{C}_c = K_c[u^{(j)}]$.) Let us introduce the following grading rk into $\widetilde{\mathcal{H}}$, $\widetilde{\mathcal{A}}^{k+}$, and $\widetilde{\mathcal{A}}^{k-}$:

(7.149) $rk(\underline{q}^{(j)}) = 1,\ rk(\underline{P}^{k(j)}) = 1,\ rk(\underline{Q}^{k(j)}) = 1,\ rk(\widetilde{C}_c) = 0.$

(7.150) $rk(v \bigtriangledown w) = rkv + rkw, \qquad rk([x, y]) = rkx + rky,$

(From Lemma 7.145 we see that (7.150) is self-consistent.)

Define, for $\alpha \in \mathbf{N}$, the following gradings:

(7.151) $\mathcal{H}_\alpha = \{x \in \mathcal{H} | rk(x) = \alpha\},$

(7.152) $\mathcal{A}^{k+}_\alpha = \{a \in \mathcal{A}^{k+} | rka = \alpha\},$

(7.153) $\mathcal{A}^{k-}_\alpha = \{e \in \mathcal{A}^{k-} | rke = \alpha\},$

with the associated filtrations

(7.154) $\mathcal{H}^\alpha = \underset{\beta \leq \alpha}{\oplus} \mathcal{H}_\beta$

(7.155) $\mathcal{A}^{\alpha k+} = \underset{\beta \leq \alpha}{\oplus} \mathcal{A}^{k+}_\beta, \qquad \mathcal{A}^{\alpha k-} = \underset{\beta \leq \alpha}{\oplus} \mathcal{A}^{k-}_\beta,$

so that

(7.156) $rk(\mathcal{H}^\alpha) \leq \alpha, rk(\mathcal{A}^{\alpha k+}) \leq \alpha, rk(\mathcal{A}^{\alpha k-}) \leq \alpha, \quad \alpha \in \mathbf{N},$

(7.157) $\underset{\alpha}{\cup} \mathcal{H}^\alpha = \mathcal{H},\ \underset{\alpha}{\cup} \mathcal{A}^{\alpha k+} = \mathcal{A}^{k+},\ \underset{\alpha}{\cup} \mathcal{A}^{\alpha k-} = \mathcal{A}^{k-},$

and analogously for $\widetilde{\mathcal{H}}^\alpha$, $\widetilde{\mathcal{H}}$, etc.

Extend naturally the invariant form (,) from $\mathcal{G}$ to $\tilde{\mathcal{H}} \subset \mathcal{G} \otimes C$. Denote by $C^\alpha, \alpha \in \mathbf{N}$, the ring (and $\mathcal{A}_0$–module) generated by the expressions

$$\text{(7.158a)} \quad \{(a,d)|a,d \in \mathcal{H},\ rk(a)+rk(d) \leq \alpha+1\},$$

$$\text{(7.158b)} \quad \left\{< x,y > |x \in \mathcal{A}^{\mathrm{k}-}, y \in \mathcal{A}^{\mathrm{k}+},\ rk(x)+rk(y) \leq \alpha+1\right\},$$

so that

$$\text{(7.159)} \quad C^\alpha \subset C^{\alpha+1}.$$

Set

$$\text{(7.160)} \quad C^\infty = \bigcup_\alpha C^\alpha,\ \tilde{C}^\alpha = C^\alpha \underset{\mathcal{A}_0}{\otimes} \tilde{C}_c,\ \tilde{C}^\infty = C^\infty \underset{\mathcal{A}_0}{\otimes} \tilde{C}_c = \bigcup_\alpha \tilde{C}^\alpha.$$

Our inductive assumption on the nature of H_r's is this:

$$\text{(7.161)} \quad H_r \in \tilde{C}^r,\quad r \in \mathbf{N}.$$

From (7.122) and (7.131) we see that this assumption is correct for $r = 1, 2$. We now have to check two things: that (7.136) can be solved, and that (7.142) yields H_{r+1} from $\tilde{C}^{r+1}$. We shall prove in §8 that

$$\text{(7.162a)} \quad \frac{\partial H}{\partial \underline{q}^{(j)}} \in \oplus(\mathcal{H}_\alpha \otimes \tilde{C}^\beta),\ \frac{\partial H}{\partial \underline{P}^{\mathrm{k}(j)}} \in \oplus(\mathcal{A}_\alpha^{\mathrm{k}+} \otimes \tilde{C}^\beta),$$

$$\frac{\partial H}{\partial \underline{Q}^{\mathrm{k}(j)}} \in \oplus(\mathcal{A}_\alpha^{\mathrm{k}-} \otimes \tilde{C}^\beta), \alpha+\beta = r, \qquad \text{for } H \in \tilde{C}^r,$$

or, equivalently,

$$\text{(7.162b)} \quad \frac{\partial H}{\partial \underline{q}^{(j)}} \in \mathcal{H} \otimes \tilde{C}^\infty,\ \frac{\partial H}{\partial \underline{P}^{\mathrm{k}(j)}} \in \mathcal{A}^{\mathrm{k}+} \otimes \tilde{C}^\infty,\ \frac{\partial H}{\partial \underline{P}^{\mathrm{k}(j)}} \in \mathcal{A}^{\mathrm{k}-} \otimes \tilde{C}^\infty,$$

for $H \in \tilde{C}^\infty$,

and that

$$\text{(7.163)} \quad -\left[\underline{q}^{(j)}, \frac{\partial H}{\partial \underline{q}^{(j)}}\right] + \frac{\partial H}{\partial \underline{P}^{\mathrm{k}(j)}} \nabla \underline{P}^{\mathrm{k}(j)} - \underline{Q}^{\mathrm{k}(j)} \nabla \frac{\partial H}{\partial \underline{Q}^{\mathrm{k}(j)}} = \mathbf{0}, \text{ for } H \in \tilde{C}^\infty.$$

In particular, (7.163) implies (7.140), so that we can find $\underline{a}_{r+1}$ from (7.136). From the results of §9, it will then follow that $D(X_{r+1})$ is symmetric. So let us analyse (7.142). From (7.162a), (7.138), (7.137), and (7.158b) we see that the last two summands in (7.142) contribute to H_{r+1} expressions belonging to $\widetilde{C}^{r+1}$, provided we assume

$$(7.164)\quad rk\, x_r \leq r,\ rk(\underline{a}_r)_i \leq r,\quad rk(\underline{y}^{\mathrm{k}}_r)_i \leq r,\quad rk(\underline{z}^{\mathrm{k}}_r)_i \leq r,\quad r \in \mathbf{N},$$

which is correct for $r = 1, 2$ (see (7.122) and (7.131)), and is recursively correct by (7.135) – (7.138). From (7.164) and (7.135a) we observe that the total derivative terms in (7.135) contribute members of $\widetilde{C}^r \subset \widetilde{C}^{r+1}$ into the first summand in (7.142), and the same conclusion applies for the terms in (7.135b) since they contribute a member of $\partial(\widetilde{C}^r)$ into $\partial(x_{r+1})$, as shows the Proof of Lemma 7.75. Now let us look at the remaining second summand in (7.142). The first term, $x_r \underline{q}$, in (7.136a), contributes, by (7.164), a member of $\widetilde{C}^{r+1}$ into $(\underline{q}, \underline{a}_{r+1})$. The remaining terms, grouped in (7.136b), can be handled similar to the way we derived (7.81):

$$-[\underline{q}, \underline{a}_r] + \underline{y}^{\mathrm{k}}_r \nabla \underline{P}^{\mathrm{k}} - \underline{Q}^{\mathrm{k}} \nabla \underline{z}^{\mathrm{k}}_r =$$

$$= -\left[\underline{q}, \frac{\delta H_r}{\delta \underline{q}}\right] + \frac{\delta H_r}{\delta \underline{P}^{\mathrm{k}}} \nabla \underline{P}^{\mathrm{k}} - \underline{Q}^{\mathrm{k}} \nabla \frac{\delta H_r}{\delta \underline{Q}^{\mathrm{k}}} =$$

$$= -\left[\underline{q}, (-\partial)^j \left(\frac{\partial H_r}{\partial \underline{q}^{(j)}}\right)\right] + (-\partial)^j \left(\frac{\partial H_r}{\partial \underline{P}^{\mathrm{k}(j)}}\right) \nabla \underline{P}^{\mathrm{k}} - \underline{Q}^{\mathrm{k}} \nabla (-\partial)^j \left(\frac{\partial H_r}{\partial \underline{Q}^{\mathrm{k}(j)}}\right)$$

$$[\text{by } (7.162\text{b})] \equiv -\left[\underline{q}^{(j)}, \frac{\partial H_r}{\partial \underline{q}^{(j)}}\right] \left(\operatorname{mod} \partial \left(\left[\underline{q}^{(j)}, \mathcal{H} \otimes \widetilde{C}^{\infty}\right]\right)\right) +$$

$$+\frac{\partial H_r}{\partial \underline{P}^{k(j)}} \nabla \underline{P}^{k(j)} \left(\mathrm{mod}\ \partial\left(\left(\mathcal{A}^{k+} \otimes \tilde{C}^{\infty}\right) \nabla \underline{P}^{k(j)}\right)\right) -$$

$$-\underline{Q}^{k(j)} \nabla \frac{\partial H_r}{\partial \underline{Q}^{k(j)}} \left(\mathrm{mod}\ \partial\left(\underline{Q}^{k(j)} \nabla \left(\mathcal{A}^{k-} \otimes \tilde{C}^{\infty}\right)\right)\right)$$

$$[\text{by } (7.163)] \equiv 0 \left(\mathrm{mod}\ \partial\left(\mathcal{H} \otimes \tilde{C}^{\infty}\right)\right).$$

(7.165)

Therefore,

$$(7.166) \quad \underline{a}_{r+1} \in \mathcal{H} \otimes \tilde{C}^{\infty},$$

and $rk\, \underline{a}_{r+1}$ is, obviously, $\leq r+1$. Hence, $(\underline{q}, \underline{a}_{r+1}) \in \tilde{C}^{r+1}$ by (7.158a), (7.166), and (7.150). Thus, $H_{r+1} \in \tilde{C}^{r+1}$, and the induction step (7.161) is completed.

To summarize, in order to establish the integrability of systems (7.61), (7.97), and (7.130), we have to fill in the following gaps: (7.80), (7.81), (7.162), (7.163), – this will be done in §8; and the supersymmetric property $D(X_{r+1})^{\dagger} = D(X_{r+1})$ for the corresponding vectors $\{X_r\}$ – this will be demonstrated in §9 in the general framework of bi-super Hamiltonian systems.

§8. Lie–Algebraic Identities

In this section we prove general results from which the formulae (7.80), (7.81), (7.162), and (7.163) follow.

In the notation of §2, let $C = \widetilde{K}[q_i^{(g|\sigma)}, \theta_\alpha^{(g|\sigma)}]$ be a commutative differential-difference superalgebra over a commutative differential-difference superalgebra $\widetilde{K}$. For any $H \in C$, define the vector-column $\frac{\partial H}{\partial \underline{q}^{(g|\sigma)}}$ by the formula

(8.1) $$\left(\frac{\partial H}{\partial \underline{q}^{(g|\sigma)}}\right)_i = \frac{\partial H}{\partial q_i^{(g|\sigma)}} .$$

To handle the first situation arising from a metrizable Lie algebra $\mathcal{G}$, suppose that the $\mathbf{Z}_2$–grading of $q_i^{(g|\sigma)}$'s are given as

(8.2) $$p(q_i^{(g|\sigma)}) = p(q_i) = p(q) \in \mathbf{Z}_2 , \quad \forall i \in I.$$

8.3. Lemma. $\frac{\partial}{\partial \underline{q}^{(g|\sigma)}}$ is a derivation of C into $C^{|I|}$, of the $\mathbf{Z}_2$ – degree $p(q)$.

Proof. Each of $\frac{\partial}{\partial q_i^{(g|\sigma)}}$'s is a derivation of C into C, having, by (8.2), the $\mathbf{Z}_2$ – degree $p(q)$. ∎

8.4. Corollary. If $v \in C$ then $v\frac{\partial}{\partial q_i^{(g|\sigma)}}$ is a derivation of C of the $\mathbf{Z}_2$–degree $p(v)+p(q)$, and $v\frac{\partial}{\partial \underline{q}^{(g|\sigma)}}$ is a derivation of C into $C^{|I|}$, of the $\mathbf{Z}_2$–degree $p(v)+p(q)$.

Let us now consider the case

(8.5) $$\widetilde{C} = \widetilde{K}[g_i^a], \quad i \in I, \quad a \in A,$$

(8.6) $$p(g_i^a) = p(a) \in \mathbf{Z}_2 , \text{ for some } \mathbf{Z}_2 \text{ –grading map } p\colon A \to \mathbf{Z}_2 .$$

Let $\mathcal{F}$ be a free n-dimensional module over $\mathcal{A}_0 \subset \widetilde{K}_0$, with a fixed basis $(e_1, ..., e_n)$ in $\mathcal{F}$. Suppose that $\mathcal{F}$ is a Lie algebra (or ring), and let $t^i_{j\mathrm{k}} \in \mathcal{A}_0$ be the structure constants of $\mathcal{F}$ in the basis $(e_1, ..., e_n)$. Let us define multiplication $[\ ,\ ]$ in

(8.7) $\widetilde{\mathcal{F}} = \mathcal{F} \underset{\mathcal{A}_0}{\otimes} \widetilde{C}$

by the rule

(8.8) $[d^1, d^2]_{\mathrm{k}} = \sum_{ij} t^{\mathrm{k}}_{ij}\, d^1_i\, d^2_j \ , \ d^1_i\, , d^2_j \in \widetilde{C}\, .$

Let us introduce the following $\mathbf{Z}_2$ – grading into $\widetilde{\mathcal{F}}$:

(8.9) $p(d_i) = p(d) \in \mathbf{Z}_2\, , \quad \forall d = (d_1, ..., d_n)^t \in \widetilde{C}^n \approx \widetilde{\mathcal{F}}\, .$

8.10. Lemma.

(8.11) $[x, y] = -(-1)^{p(x)p(y)}\, [y, x]\, , \quad \forall x, y \in \widetilde{\mathcal{F}}\, .$

Proof. Formulae (8.8)and (8.9). ∎

8.12. Lemma. For any $x, y, z \in \widetilde{\mathcal{F}}$,

(8.13) $[[x, y]\, , z] = [x,\ [y, z]] - (-1)^{p(x)p(y)}\, [y, [x, z]]\, .$

Proof. We have

$([[x, y]\, , z] - [x, [y, z]])_{\mathrm{k}} = t^{\mathrm{k}}_{ij}\, ([x, y]_i\, z_j - x_i[y, z]_j) =$

$= t^{\mathrm{k}}_{ij}\left(t^{i}_{\alpha\beta}\, x_\alpha\, y_\beta\, z_j - t^{j}_{\alpha\beta}\, x_i\, y_\alpha\, z_\beta\right) = x_\alpha\, y_\beta\, z_\gamma\left(t^{\mathrm{k}}_{i\gamma}\, t^{i}_{\alpha\beta} - t^{\mathrm{k}}_{\alpha j}\, t^{j}_{\beta\gamma}\right)$ [since $\mathcal{F}$ is a Lie algebra] $= x_\alpha\, y_\beta\, z_\gamma\left(-t^{\mathrm{k}}_{\beta i}\, t^{i}_{\alpha\gamma}\right) = -(-1)^{p(x)p(y)}\, t^{\mathrm{k}}_{\beta i}\, t^{\mathrm{k}}_{\alpha\beta}\, y_\beta\, x_\alpha\, z_\gamma =$

$= -(-1)^{p(x)p(y)} t^{\mathrm{k}}_{\beta i} y_\beta [x, z]_i = -(-1)^{p(x)p(y)}([y, [x, z]])_{\mathrm{k}}\ .$ ∎

8.14. Remark. We shall call $\widetilde{\mathcal{F}}$ a *superLie algebra.* The reader is asked to look up Definition 3.72 of a *Lie superalgebra* to appreciate that although we have the skewsymmetry (8.11) and the graded Jacobi identity (8.13) in $\tilde{\mathcal{F}}$, $\tilde{\mathcal{F}}$ is *not* a Lie superalgebra: the rule (8.8) is not of the type (3.73). The reason for adopting formula (8.8) for the commutator in $\tilde{\mathcal{F}}$ is that formula (7.71) is precisely of this type; hence, the relation (7.77) we are after has the same meaning of the commutator as the one we have adopted in (8.8).

Suppose now that

$$(8.15)\quad (x,y) = \sum_{i=1}^{n} x_i\, y_i$$

is an invariant scalar product in $\mathcal{F}$:

$$(8.16)\quad ([x,y],z) = (x,[y,z])\ ,\quad \forall x,y,z \in \mathcal{F}\ .$$

8.17. Lemma.

$$(8.18)\quad ([x,y],z) = (x,[y,z])\ ,\quad \forall x,y,z \in \tilde{\mathcal{F}}\ .$$

Proof. The identity (8.16) written as

$$(8.19)\quad t^i_{jk}\, x_j\, y_k\, z_i = x_j\, t^j_{ki}\, y_k\, z_i\ ,\quad \forall x,\ y,\ z\ \in \mathcal{F}\ ,$$

is equivalent to the structure constants t^i_{jk} being cyclic-symmetric :

$$(8.20)\quad t^i_{jk} = t^j_{ki}\ ,$$

and this implies (8.18) as well. ∎

8.21. Lemma.

$$(8.22)\quad (x,y) = (-1)^{p(x)p(y)}\,(y,x)\ ,\quad \forall x,y \in \tilde{\mathcal{F}}\ .$$

Proof. Formulae (8.15) and (8.9). ∎

Denote by $\mathcal{F}^\wedge = \mathcal{F}^\wedge_g$ the superLie algebra formed from the vectors $\{g^a = (g^a_i), i \in I = \{1,...,n\}\,,\ a \in A\}$ by the commutator rule (8.18). We have

(8.23) $\mathcal{F}^\wedge \subset \mathcal{F} \underset{\mathcal{A}_0}{\otimes} \mathcal{A}_0 [g_i^a]$,

(8.24) $\mathcal{F}^\wedge(\widetilde{K}) := \mathcal{F}^\wedge \underset{\mathcal{A}_0}{\otimes} \widetilde{K} \subset \mathcal{F} \underset{\mathcal{A}_0}{\otimes} \widetilde{K}\,[g_i^a] = \widetilde{\mathcal{F}}$.

Set

(8.25) $\mathcal{R}^\wedge = \mathcal{R}_g^\wedge =$ the $\mathcal{A}_0$−module and the ring generated by

$\{(x,y) \big| x, y \in \mathcal{F}^\wedge\}$.

Then

(8.26) $\mathcal{R}^\wedge(\widetilde{K}) := \mathcal{R}^\wedge \underset{\mathcal{A}_0}{\otimes} \widetilde{K} =$ the $\widetilde{K}$−module and the ring generated by

$\{(x,y) \big| x, y \in \mathcal{F}^\wedge(\widetilde{K})\}$.

Notice that $\mathcal{R}^\wedge \subset \mathcal{A}_0\,[g_i^a]$ and $\mathcal{R}^\wedge(\widetilde{K}) \subset \widetilde{K}\,[g_i^a] = \widetilde{C}$.

Now we can formulate the first main result of this section.

8.27. Theorem.

(8.28) $\sum_a \left[g^a, \frac{\partial H}{\partial g^a} \right] = 0\,, \quad \forall H \in \mathcal{R}^\wedge(\widetilde{K})$,

where $\frac{\partial H}{\partial g^a}$ denotes the vector $\left(\frac{\partial H}{\partial g_i^a} \right)$.

8.29. Corollary. The third equality in (7.81) is justified when we specialize our situation by setting:

(8.30) $\widetilde{K} = K_c[u^{(j)}, \varphi^{(j)}]$, $\{g_i^a\} = \{q_i^{(j)}, \sigma_i^{(j)}\}$, $\mathcal{F} = \mathcal{G}$.

This yields the identifications

(8.31) $\mathcal{F}^\wedge = \widehat{\mathcal{G}}\,, \mathcal{R}^\wedge = \mathcal{M}_{\widehat{g}}\,, \; \mathcal{R}^\wedge(\widetilde{K}) = \mathcal{R}$.

In particular, (7.77) holds true for $H_r \in \mathcal{R}$.

8.32. Remark. We could have taken $\widetilde{K} = K[u^{(j)}, \varphi^{(j)}]$ instead of

$\widetilde{K} = K_c[u^{(j)}, \varphi^{(j)}]$ in (8.30), leaving (7.77) still satisfied: K_c comes from Lemma 7.75.

We break the Proof of Theorem 8.27 into a sequence of simple steps.

8.33. Lemma. The map $\mathcal{R}^\wedge(\widetilde{K}) \hookrightarrow \widetilde{C} \to \widetilde{C}^n$, given by the L.H.S. of (8.28), is an even derivation over $\widetilde{K}$.

Proof. By (8.8),

$$\left[g^a, \frac{\partial H}{\partial g^a}\right]_{\mathrm{k}} = t^{\mathrm{k}}_{ij}\, g^a_i \frac{\partial H}{\partial g^a_j},$$

which is an even derivation by Corollary 8.4 and (8.6). ∎

8.34. Corollary. It is enough to check (8.28) for H's of the form

(8.35) $H = (x, y), \ \ x, y \in \mathcal{F}^\wedge$.

8.36. Lemma. Let $x^\omega = (x^\omega_i), \omega \in \Omega$, be another set of variables, with the $\mathbf{Z}_2$-gradings

(8.37) $p(x^\omega_i) = p(\omega) \in \mathbf{Z}_2$.

Form $\mathcal{F}^\wedge_x, \mathcal{R}^\wedge_x$, etc. Pick any $h \in \mathcal{R}^\wedge_x(\widetilde{K})$; say, $h = h(x^\alpha, x^\beta, ..)$, $\alpha, \beta, ... \in \Omega_1 \subset \Omega$. If $H \in \mathcal{R}^\wedge_g(\widetilde{K})$ is such that

(8.38) $H = h\big|_{x^\alpha = g^{f(\alpha)}, \ldots}$

for some map $f: \Omega_1 \to A$ commuting with the grading maps $p: (\cdot) \to \mathbf{Z}_2$, then

(8.39) $$\left[g^a, \frac{\partial H}{\partial g^a}\right] = \left[x^\omega, \frac{\partial h}{\partial x^\omega}\right]\Bigg|_{x^\alpha = g^{f(\alpha)}, \ldots}$$

8.40. Remark. The representation (8.38), which I call the Shapovalov trick, plays the role of polarization.

Proof. Extend f by defining $f: (\Omega \backslash \Omega_1) \to A$ arbitrarily (but commuting with the $\mathbf{Z}_2$–gradings). We have

$$(8.41)\quad \frac{\partial H}{\partial g^a} = \sum_{\omega} \frac{\partial h}{\partial x^{\omega}} \, \delta_a^{f(\omega)} \Big|_{x^{\alpha}=g^{f(\alpha)},\ldots}$$

Therefore,

$$\sum_a \left[g^a, \frac{\partial H}{\partial g^a} \right] = \sum_{a,\omega} \left[g^a, \frac{\partial h}{\partial x^{\omega}} \delta_a^{f(\omega)} \right] \Big|_{x^{\alpha}=g^{f(\alpha)},\ldots} =$$

$$= \sum_{\omega} \left[x^{\omega}, \frac{\partial h}{\partial x^{\omega}} \right] \Big|_{x^{\alpha}=g^{f(\alpha)},\ldots}$$ ∎

8.42. Corollary. It is enough to check (8.28) for H's of the form

(8.43) $H = (x, y)$, $x, y \in \mathcal{F}^{\wedge}$, H is polylinear in g's, provided we enlarge A into $A \cup A \cup \ldots$, which we assume is done from now on.

Denote by Plin the set of such H's polylinear in g's (with A enlarged as agreed).

8.44. Lemma. Suppose $H \in$ Plin , and

(8.45) $H = (g^a, x)$, $x \in \mathcal{F}^{\wedge}$.

Then

$$(8.46)\quad \frac{\partial H}{\partial g^a} = x \, .$$

Proof. Since $H \in$ Plin , x does not contain g^a. By(8.15) , $H = g_i^a x_i$.

Hence $\left(\frac{\partial H}{\partial g^a} \right)_i = \frac{\partial H}{\partial g_i^a} = x_i$. ∎

8.47. Corollary. Property (7.80) is justified.

Proof follows from (8.41), (8.46), (8.18), and Corollary 8.4, specialized by (8.30), (8.31). ∎

Proof of Theorem 8.27. By Corollary (8.42), it is enough to check

(8.28) for H's of the form (8.43). We use induction on the number l of different g's involved in H. First, for $l = 2$, if

(8.48) $H = (g^a, g^b)\,, a \neq b\,,$

then, by Lemma 8.44,

(8.49) $\dfrac{\partial H}{\partial g^a} = g^b,$

and by (8.22) and Lemma 8.44,

(8.50) $\dfrac{\partial H}{\partial g^b} = (-1)^{p(a)p(b)} g^a.$

Hence,

$$\sum_{c=a,b} \left[g^c, \frac{\partial H}{\partial g^c}\right] = [g^a, g^b] + \left[g^b, (-1)^{p(a)p(b)} g^a\right] = 0$$

by (8.11). Suppose now that we have verified (8.28) for all H in (8.43) involving $l \leq L+1$ different g's. Notice that from the Jacobi identity (8.13), skewsymmetry (8.11), and the invariance property (8.18), it follows that Plin is generated over **Z** by the elements of the form

(8.51) $(x, g^a) \in$ Plin.

It is also generated by the elements of the form

(8.52a) $(g^b, g^a)\,, a \neq b\,,$

and

(8.52b) $([x, g^b], g^a) \in$ Plin .

We now show that if (8.51) satisfies (8.28) then so does (8.52b). Suppose $H = (x, g^a) \in$ Plin, and x involves L different g's, denoted by $x_1, \ldots, x_L$. Then, by the inductive assumption,

$$(8.53)\quad 0=\sum_c\left[g^c,\frac{\partial H}{\partial g^c}\right]=\sum_{j=1}^{L}\left[x_j,\frac{\partial H}{\partial x_j}\right]+\left[g^a,\frac{\partial H}{\partial g^a}\right].$$

Hence, by (8.46) and (8.22),

$$(8.53)\quad \sum_{j=1}^{L}\left[x_j,\frac{\partial(x,g^a)}{\partial x_j}\right]=-\left[g^a,(-1)^{p(a)p(x)}x\right]=[x,g^a],$$

which implies that

$$(8.54)\quad \sum_{j=1}^{L}\left[x_j,\frac{\partial(x,y)}{\partial x_j}\right]=[x,y],$$

$\forall(x,y)\in$ Plin, that is, provided y does not involve $x_1,\ldots,x_L$. Therefore, for

$H=([x,g^b],g^a)=(x,[g^b,g^a])=(-1)^{p(x)p(b)+1}(g^b,[x,g^a])$ in (8.52b) we have,

$$\sum\left[g^c,\frac{\partial H}{\partial g^c}\right]=$$

$$(8.55)\quad \sum_{j=1}^{L}\left[x_j,\frac{\partial(x,[g^b,g^a])}{\partial x_j}\right]+$$

$$(8.56a)\quad \left[g^b,(-1)^{p(x)p(b)+1}[x,g^a]\right]+$$

$$(8.56b)\quad [g^a,(-1)^{p(a)[p(x)+p(b)]}[x,g^b]].$$

By (8.54), (8.55) equals to

$$(8.56c)\quad [x,[g^b,g^a]],$$

and all three terms in (8.56) add up to zero by the graded Jacobi identity (8.13). ∎

We now turn to the second situation arising from a set of representations $\rho_k\colon \mathcal{F}\to End\ V_k$. Mercifully, now we do not have to worry about $\mathbf{Z}_2$−gradings since we are dealing with a purely even situation. We set

(8.57) $\widetilde{C} = \widetilde{K}\left[g_i^a, P_{j(k)}^{ks}, Q_{j(k)}^{ks}\right], a \in A, k \in A_1, s \in S, 1 \leq i \leq n =$
$= \dim \mathcal{F}, 1 \leq j(k) \leq n_k = \dim V_k,$

where $\widetilde{K} \supset \mathcal{A}_0$ is now a commutative algebra. Denote

(8.58) $< w, v >= \delta_\ell^k \sum_{j=1}^{n_k} w_j v_j$ for $w \in \widetilde{V}_k^* : = V_k^* \underset{\mathcal{A}_0}{\otimes} \widetilde{C}, v \in \widetilde{V}_\ell : = V_\ell \underset{\mathcal{A}_0}{\otimes} \widetilde{C},$

and let (8.15) be an invariant scalar product in both $\mathcal{F}$ and $\widetilde{\mathcal{F}} = \mathcal{F} \otimes \widetilde{C}$.

Denote by $\nabla: \widetilde{V}_\ell \times \widetilde{V}_k^* \to \widetilde{\mathcal{F}}$ the map defined by the equation

(8.59) $(v \nabla w, a) =< w, a.v >, v \in \widetilde{V}_\ell, w \in \widetilde{V}_k^*, a \in \widetilde{\mathcal{F}},$

where $a.v$ stands for $\rho_\ell(a)(v)$, and ρ_k's (resp., $\rho_k^\dagger$'s) are understood to be naturally extended into the representations of $\widetilde{\mathcal{F}}$ in $End\, \widetilde{V}_k$ (resp., $End\, \widetilde{V}_k^*$).

Denote by

(8.60) $g^a = (g_i^a), \quad P^{ks} = \left(P_{j(k)}^{ks}\right), Q^{ks} = \left(Q_{j(k)}^{ks}\right)$

the corresponding column-vectors. Define the following $\mathcal{A}_0$-modules

(8.61) $\mathcal{F}_1 = \underset{a}{\oplus} \mathcal{A}_0\, g^a \subset \widetilde{\mathcal{F}},$

(8.62) $\mathcal{A}_1^{k+} = \underset{s}{\oplus} \mathcal{A}_0 Q^{ks} \subset \widetilde{V}_k, \mathcal{A}_1^{k-} = \underset{s}{\oplus} \mathcal{A}_0\; P^{ks} \subset \widetilde{V}_k^*.$

Next, recursively, for each $r \in \mathbf{N} + 1$ define $\mathcal{A}_0$-modules generated as follows

(8.63) $\mathcal{F}_r$: $\{[\mathcal{F}_\alpha, \mathcal{F}_\beta] | \alpha + \beta = r\}$ and $\left\{\mathcal{A}_\alpha^{k+} \nabla \mathcal{A}_\beta^{k-} | \alpha + \beta = r, \text{ no sum on } k\right\},$

(8.64) $\mathcal{A}_r^{k+}$: $\{\mathcal{F}_\alpha \cdot \mathcal{A}_\beta^{k+} | \alpha + \beta = r\}, \mathcal{A}_r^{k-}$: $\{\mathcal{F}_\alpha \cdot \mathcal{A}_\beta^{k-} | \alpha + \beta = r\}.$

Set

(8.65) $\mathcal{F}^\vee = \underset{r \in \mathbf{N}}{\oplus} \mathcal{F}_r, \mathcal{A}^{k+} = \underset{r \in \mathbf{N}}{\oplus} \mathcal{A}_r^{k+}, \mathcal{A}^{k-} = \underset{r \in \mathbf{N}}{\oplus} \mathcal{A}_r^{k-},$

and define

$$(8.66)\quad \mathcal{F}_r(\widetilde{K}) = \mathcal{F}_r \underset{\mathcal{A}_0}{\otimes} \widetilde{K}\,, \mathcal{F}^\vee(\widetilde{K}) = \mathcal{F}^\vee \otimes \widetilde{K} \subset \widetilde{\mathcal{F}}\,,$$

and analogously for $\mathcal{A}_r^{\mathbf{k}+}$, etc. By construction, $\mathcal{F}^\vee$ is a Lie algebra (over $\mathcal{A}_0$), $\mathcal{A}^{\mathbf{k}+}$ and $\mathcal{A}^{\mathbf{k}-}$ are $\mathcal{A}_0-$ and $\mathcal{F}^\vee-$ modules, for each k there is a map $\nabla\colon \mathcal{A}^{\mathbf{k}+} \times \mathcal{A}^{\mathbf{k}-} \to \mathcal{F}^\vee$, and the set $\{\mathcal{F}^\vee, \mathcal{A}^{\mathbf{k}+}, \mathcal{A}^{\mathbf{k}-}\}$ is the minimal set with these properties satisfying the "Cauchy data"(8.61) , (8.62). Let

(8.67) $\mathcal{N} =$ the $\mathcal{A}_0$–module and the ring generated by the expressions $\{(a,b)\big|a,b \in \mathcal{F}^\vee\}$ and $\{< v,u > \big|u \in \mathcal{A}^{\mathbf{k}+}, v \in \mathcal{A}^{\mathbf{k}-}\}$.

Noticing that, thanks to Lemma 7.145, the subscripts in (8.63), (8.64) can be taken to define gradings on $\mathcal{F}^\vee$, $\mathcal{F}^\vee(\widetilde{K})$, etc., – let us denote it rk, – we can induce this gradation onto $\mathcal{N}$:

$$(8.68)\quad \mathcal{N} = \bigoplus_{r \in \mathbf{N}+1} \mathcal{N}_r,$$

where we naturally set

$$(8.69)\quad rk(a,b) = rka + rkb, rk < v,u >= rkv + rku, rk\,\mathcal{A}_0 = 0\,.$$

Extend this gradation onto $\mathcal{N}(\widetilde{K}) = \mathcal{N} \underset{\mathcal{A}_0}{\otimes} \widetilde{K}$ by letting

$$(8.70)\quad rk(\widetilde{K}) = 0\,.$$

The second main result of this section is

8.71. Theorem.

$$(8.72)\quad \left[g^a, \frac{\partial H}{\partial g^a}\right] - \frac{\partial H}{\partial P^{\mathbf{k}_s}} \nabla P^{\mathbf{k}_s} + Q^{\mathbf{k}_s} \nabla \frac{\partial H}{\partial Q^{\mathbf{k}_s}} = 0\,, \forall H \in \mathcal{N}(\widetilde{K})\,.$$

8.73. <u>Corollary.</u> The formula (7.163) is proved when we specialize our set-up to the case

$$\widetilde{K} = K_c[u^{(j)}] = \widetilde{C}_c\,, \{g_i^a\} = \{q_i^{(j)}\}, \{P_{j(\mathrm{k})}^{\mathrm{k}_s}\} = \{P_\alpha^{\mathrm{k}(j)}\}, \{Q_{j(\mathrm{k})}^{\mathrm{k}_s}\} = \{Q_\alpha^{\mathrm{k}(j)}\},$$

(8.74) $\mathcal{F} = \mathcal{G}, \mathcal{F}_r = \mathcal{H}_r,\ \mathcal{A}_r^{\mathrm{k}+} = \mathcal{A}_r^{\mathrm{k}+}, \mathcal{A}_r^{\mathrm{k}-} = \mathcal{A}_r^{\mathrm{k}-}, \mathcal{F}^\vee = \mathcal{H},$

$$\mathcal{A}^{\mathrm{k}+} = \mathcal{A}^{\mathrm{k}+}, \mathcal{A}^{\mathrm{k}-} = \mathcal{A}^{\mathrm{k}-}, \bigoplus_{r \le \alpha+1} \qquad \mathcal{N}_r = C^\alpha\,, \mathcal{N} = C^\infty\,, \mathcal{N}(\widetilde{K}) = \widetilde{C}^\infty\,.$$

8.75. Remark. Theorem 8.71 is a nontrivial generalization of the purely even case of Theorem 8.27.

We break the Proof of Theorem 8.71 into a sequence of simple steps similar to the ones we used to prove Theorem 8.27.

8.76. Lemma. The map $\mathcal{N}(\widetilde{K}) \hookrightarrow \widetilde{C} \to \widetilde{C}^{\,n}$ given by the L.H.S. of (8.72), is a derivation.

Proof is obvious. ∎

8.77. Corollary. It is enough to check (8.72) for H's of the form

(8.78) $\{(a,b) | a, b \in \mathcal{F}^\vee\}, \quad \{< v, u > | a \in \mathcal{A}^{\mathrm{k}+}, v \in \mathcal{A}^{\mathrm{k}-}\}$

8.79. Lemma. Suppose that, like in Lemma 8.36, $\{x^\omega,\, y^\mu,\, z^\mu\}$ is a new set of variables. Attach the superscript ``prime´to the notation of the corresponding objects, like $\mathcal{N}'$, etc. Pick any $h \in \mathcal{N}(\widetilde{K})'$, say,

(8.80) $h = h(x^\alpha,\, y^\mu,\, z^\mu, \ldots).$

If $H \in \mathcal{N}(\widetilde{K})$ is such that

(8.81) $H = h|_{x^\alpha = g^{f_1(\alpha)},\ y^\mu = P^{f_2(\mu)},\ z^\mu = Q^{f_3(\mu)}, \ldots}$

for some maps $f_1 : \Omega_1 \to A$, etc., then

(8.82) $\left[g^a, \dfrac{\partial H}{\partial g^a}\right] - \dfrac{\partial H}{\partial P^{\mathrm{k}_s}} \bigtriangledown P^{\mathrm{k}_s} + Q^{\mathrm{k}_s} \bigtriangledown \dfrac{\partial H}{\partial Q^{\mathrm{k}_s}} =$

$$= \left(\left[x^{\omega}, \frac{\partial h}{\partial x^{\omega}} \right] - \frac{\partial h}{\partial y^{\mu}} \nabla y^{\mu} + z^{\mu} \nabla \frac{\partial h}{\partial z^{\mu}} \right) \Big|_{x^{\alpha} = g^{f_1(\alpha)}, \dots} .$$

Proof. We have, extending the maps f_1, etc.,:

$$\text{(8.83a)} \quad \frac{\partial H}{\partial g^a} = \sum_{\omega} \frac{\partial h}{\partial x^{\omega}} \delta_a^{f_1(\omega)} \Big|_{x^{\alpha} = g^{f_1(\alpha)}, \dots},$$

$$\text{(8.83b)} \quad \frac{\partial H}{\partial P^{\mathrm{k}_s}} = \sum_{\mu} \frac{\partial h}{\partial y^{\mu}} \delta_{\mathrm{k}_s}^{f_2(\mu)} \Big|_{x^{\alpha} = g^{f_1(\alpha)}, \dots},$$

$$\text{(8.83c)} \quad \frac{\partial H}{\partial Q^{\mathrm{k}_s}} = \sum_{\mu} \frac{\partial h}{\partial z^{\mu}} \delta_{\mathrm{k}_s}^{f_3(\mu)} \Big|_{x^{\alpha} = g^{f_1(\alpha)}, \dots},$$

and the same argument as in the Proof of Lemma 8.36, yields (8.82). ∎

8.84. Corollary. It is enough to check (8.72) for only those H's of the form (8.78) which are polylinear in g's, P's, and Q's, provided we enlarge A, A_1, and S into $A \cup A \cup \dots$, etc., which we assume is done henceforth.

Denote by Plin the set of such polylinear H's.

Proof of Theorem 8.71. By Corollary 8.84, it is enough to check (8.72) for $H \in$ Plin. We shall do this by induction on $rk\,H$. First, if $rk\,H = 2$ then H is of one of the two forms

$$\text{(8.85)} \quad H = (g^a, g^b), \quad a \neq b,$$

$$\text{(8.86)} \quad H = \langle Q^{\mathrm{k}_s}, P^{\mathrm{k}_r} \rangle .$$

The case (8.85) requires no checking being under the auspicies of Theorem 8.27. For H in (8.86), we have $H = \sum_{j=1}^{n_{\mathrm{k}}} Q_j^{\mathrm{k}_s} P_j^{\mathrm{k}_r}$ by (8.58), so that

$$\text{(8.87)} \quad \frac{\partial H}{\partial P^{\mathrm{k}_r}} = Q^{\mathrm{k}_s}, \quad \frac{\partial H}{\partial Q^{\mathrm{k}_s}} = P^{\mathrm{k}_r},$$

and the L.H.S. of (8.72) becomes

$$-Q^{\mathrm{k}_s} \nabla P^{\mathrm{k}_r} + Q^{\mathrm{k}_s} \nabla P^{\mathrm{k}_r} = 0,$$

as desired. Let now $rk\, H = 3$. (The reason why we need to base our induction on *both* $rk = 2-$ and $rk = 3-$ cases, will become clear later on.) To see what kind of H's we can have, let us first list generators of $\mathcal{F}_2$, $\mathcal{A}_2^{\mathrm{k}+}$ and $\mathcal{A}_2^{\mathrm{k}-}$, by using (8.63) and (8.64):

$$(8.88) \quad \mathcal{F}_2 : \quad [g^a, g^b];\ Q^{\ell s} \bigtriangledown P^{\ell r} \text{ (no sum on } \ell);$$

$$(8.89) \quad \mathcal{A}_2^{\mathrm{k}+} : \quad g^a.Q^{\mathrm{k}s};$$

$$(8.90) \quad \mathcal{A}_2^{\mathrm{k}-} : \quad g^a.P^{\mathrm{k}s}.$$

Hence, elements of Plin of $rk = 3$ are generated by

$$(8.91a) \quad ([g^a, g^b], g^c), \quad a \neq b \neq c \neq a,$$

$$(8.91b) \quad (g^a,\ Q^{\ell s} \bigtriangledown P^{\ell r}) \text{ (no sum on } \ell),$$

$$(8.91c) \quad < g^a.P^{\ell r}, Q^{\ell s} >; \quad < P^{\ell r}, g^a.Q^{\ell s} > \text{ (no sum on } \ell).$$

The case (8.91a) requires no checking since it falls under the authority of Theorem 8.27, while either of the two expressions in (8.91c) reduces, by (8.59) and (7.99), to the type (8.91b). So, let

$$(8.92)\ H = (g^a, Q^{\ell s} \bigtriangledown P^{\ell r}) =< P^{\ell r}, g^a.Q^{\ell s} >= - < g^a.P^{\ell r}, Q^{\ell s} > \text{ (no sum on } \ell)$$

Then, by (8.15) and (8.58),

$$(8.93) \quad \frac{\partial H}{\partial g^a} = Q^{\ell s} \bigtriangledown P^{\ell s}, \frac{\partial H}{\partial P^{\ell r}} = g^a.Q^{\ell s}, \frac{\partial H}{\partial Q^{\ell s}} = -g^a.P^{\ell r},$$

and the L.H.S. of (8.72) becomes

$$[g^a, Q^{\ell s} \bigtriangledown P^{\ell s}] - (g^a.Q^{\ell s}) \bigtriangledown P^{\ell r} - Q^{\ell s} \bigtriangledown (g^a.P^{\ell r}),$$

which is zero by Lemma 7.145. Thus, the $rk = 3$–case is verified.

Now, suppose we have checked (8.72) for all $H \in$ Plin with $rk\, H \leq \ell + 1$. Notice that by (8.64), every $H \in$ Plin of the form $H =< v, u >$, $u \in \mathcal{A}_\alpha^{\mathrm{k}+}$, $v \in$

$\mathcal{A}_{\beta}^{\mathrm{k}-}$, with $rk\, H = \alpha + \beta > 2$, may be also written in the form

(8.94) $H = (a, b) \in \text{Plin}, \; a \in \mathcal{F}_{\gamma}, \; b \in \mathcal{F}_{\delta}, \; \gamma + \delta = rk\, H,$

by use of (8.59). So, suppose (8.72) is satisfied for all H of the form

(8.95) $H = (a, g^c) \in \text{Plin}, \; a \in \mathcal{F}_{\alpha}, \; \alpha \leq \ell.$

Let $\{x\} = \{x_1, \ldots, x_{\alpha}\}$ denote the set of all the α different (vector) variables entering into a in (8.95). Denote the operator acting on $H \in$ Plin in the L.H.S. of (8.72) as $\mathcal{O}_y$ where $\{y\}$ denotes the set of all the different variables in $H \in$ Plin. Then (8.72) for our H from (8.95) can be written in the form

$$0 = \underset{x \in a}{\mathcal{O}_x} \; ((a, g^c)) - [g^c, a],$$

or

$$\underset{x \in a}{\mathcal{O}_x} \; ((a, g^c)) = [a, g^c],$$

which implies

(8.96) $\underset{x \in a}{\mathcal{O}_x} \; ((a, b)) = [a, b], \quad \forall a \in \mathcal{F}_{\alpha}, \; \alpha \leq \ell, \; \forall b \in \mathcal{F}^{\vee}, \; b \not\ni x \;$ for $x \in a$.

In particular,

(8.97) $\underset{y \in b}{\mathcal{O}_y} \; ((b, a)) = [b, a], \quad \forall b \in \mathcal{F}_{\beta}, \; \beta \leq \ell, \; \forall a \in \mathcal{F}^{\vee}, \; a \not\ni y \;$ for $y \in b$.

Adding (8.96) to (8.97), we find that

$$(8.98) \quad \begin{cases} (8.72) \text{ is satisfied for all } H = (a, b) \in \text{Plin, with } rk\, a \leq \ell, \; rk\, b \leq \ell, \\ \text{provided it is satisfied for all } H = (a, g^c) \in \text{Plin, with } rk\, a \leq \ell. \end{cases}$$

To complete the induction step, it remains to show that (8.72) is also satisfied for

(8.99) $H = (a, g^c) \in \text{Plin}, \quad a \in \mathcal{F}_{\ell+1}.$

From (8.63) we see that there are two types of possibilities to consider:

(8.100.1) $a = [\varphi, \psi], \quad rk\,\varphi \geq rk\,\psi, \quad rk\,\varphi + rk\,\psi = \ell + 1,$

(8.100.2) $a = \varphi \bigtriangledown \psi, \quad \varphi \in \mathcal{A}^{\mathrm{k}+}, \; \varphi \in \mathcal{A}^{\mathrm{k}-}, \quad rk\,\varphi + rk\,\psi = \ell + 1.$

We consider the case (8.100.1) first. Here $H = (a, g^c) = ([\varphi, \psi], g^c) = (\varphi, [\psi, g^c])$, and $rk\,\varphi = \ell + 1 - rk\psi \leq \ell$, $rk\,([\psi, g^c]) = rk\,\psi + 1 \leq \dfrac{\ell+1}{2} + 1 = \dfrac{\ell+3}{2} \leq \ell$ for $\ell \geq 3$. Now, we have covered ourselves by checking the cases $\ell = 1, 2$ as the basis of our induction. If $\ell = 3$, then $rk\,\varphi = 2$, $rk\,\psi = 1$, and $rk\,([\psi, g^c]) = 2$, so that we can use (8.98) to establish the validity of (8.72) for $([\mathcal{F}_2, \mathcal{F}_1], \mathcal{F}_1)$. To finish, we need to complete the $(\mathcal{F}_3, \mathcal{F}_1)$-situation by considering the case of $(\varphi \bigtriangledown \psi, \, g^c)$, $rk\,\varphi + rk\,\psi = 3$. We shall use the identity

(8.101) $(\varphi \bigtriangledown \psi, \, g^c) = \, < \psi, g^c.\varphi > \, = \, - < g^c.\psi, \varphi >$

If $rk\,\varphi = 2, \quad rk\,\psi = 1, \quad$ say, $\varphi = g^a.Q^{\ell s}, \quad \psi = P^{\ell r}, \quad$ then

(8.102) $(\varphi \bigtriangledown \psi, g^c) = \, - < g^c.P^{\ell r}, \; g^a.Q^{\ell s} > .$

If, on the contrary, $rk\,\varphi = 1$, $rk\,\psi = 2$, say, $\varphi = Q^{\ell s}$, $\psi = g^a.P^{\ell r}$, then

(8.103) $(\varphi \bigtriangledown \psi, g^c) = \, < g^a.P^{\ell r}, \; g^c.Q^{\ell s} >,$

which is of the same form as (8.102). Now, for H of the form (8.102), we have

$$H = (\varphi \bigtriangledown \psi, g^c) = ((g^a.Q^{\ell s}) \bigtriangledown P^{\ell r}, g^c) = \, - < g^c.P^{\ell r}, \; g^a.Q^{\ell s} > =$$

$$= \, < P^{\ell r}, \, g^c.(g^a.Q^{\ell s}) > = \, < g^a.(g^c.P^{\ell r}), \, Q^{\ell s} > = -(Q^{\ell s} \bigtriangledown (g^c.P^{\ell r}), g^a).$$

Therefore,

$$\frac{\partial H}{\partial g^c} = (g^a.Q^{\ell s}) \bigtriangledown P^{\ell r}, \quad \frac{\partial H}{\partial g^a} = -Q^{\ell s} \bigtriangledown (g^c.P^{\ell r}),$$
$$\frac{\partial H}{\partial P^{\ell r}} = g^c.(g^a.Q^{\ell s}), \quad \frac{\partial H}{\partial Q^{\ell s}} = g^a.(g^c.P^{\ell r}),$$

and the L.H.S. of (8.72) becomes

$$[g^c, (g^a.Q^{\ell s}) \bigtriangledown P^{\ell r}] - [g^a, Q^{\ell s} \bigtriangledown (g^c.P^{\ell r})] -$$
$$-(g^c.(g^a.Q^{\ell s})) \bigtriangledown P^{\ell r} + Q^{\ell s} \bigtriangledown (g^a.(g^c.P^{\ell r}))$$

which adds up to zero by Lemma 7.145. Thus the case (8.100.1) is disposed off.

Next, the case (8.100.2). Write

$$\varphi = x_1.(x_2.(\dots .(x_p.Q^{\ell s})\dots),$$
$$(8.104) \qquad x_i, y_j \in \mathcal{F}^\vee.$$
$$\psi = y_1.(y_2(\dots .(y_q.P^{\ell r})\dots),$$

If amongst x's and y's there is at least one element of $rk > 1$, say z, we can rewrite H in (8.99) in the form $H = (b, z) \in$ Plin, with $rk\, b \leq \ell$, $rk\, z = \ell + 1 - rk\, b \leq \ell$, by the repeated use of (8.101), (8.59), and (7.99), and this case is covered by (8.98). Therefore, it remains to consider the case of (8.104) with all the x's and y's being various g's. We have,

$$H = (x_1.(\dots .(x_p.Q^{\ell s})\dots) \bigtriangledown (y_1.(\dots .(y_q.P^{\ell r})\dots), g^a) =$$
$$=< y_1.(\dots .(y_q.P^{\ell r})\dots), g^a.(x_1.(\dots .(x_p.Q^{\ell s})\dots) >=$$
$$= (-1)^q < P^{\ell r}, y_q.(\dots .(y_1.(g^a.(x_1.(\dots .(x_p.Q^{\ell s})\dots) > .$$

At this point it is convenient to change notation slightly by denoting: $P^{\ell r}$ by P; $Q^{\ell s}$ by Q; $y_q, \dots, y_1, g^a, x_1, \dots, x_p$ by $z_1, \dots, z_\ell$. Thus,

$$H =< P, z_1.(\dots .(z_\ell.Q)\dots) >= (-1)^\ell < z_\ell.(\dots .(z_1.P)\dots), Q >=$$

$$= (-1)^{r-1}((z_{r+1}.(\dots .(z_\ell.Q)\dots) \nabla (z_{r-1}.(\dots .(z_1.P)\dots), z_r).$$

Therefore,

$$\frac{\partial H}{\partial z_r} = (-1)^{r-1}(z_{r+1}.(\dots .(z_\ell.Q)\dots) \nabla (z_{r-1}.(\dots .(z_1.P)\dots),$$

$$\frac{\partial H}{\partial P} = z_1.(\dots .(z_\ell.Q)\dots),$$

$$\frac{\partial H}{\partial Q} = (-1)^\ell z_\ell.(\dots .(z_1.P)\dots).$$

Hence,

$$\left[z_r, \frac{\partial H}{\partial z_r}\right] - \frac{\partial H}{\partial P} \nabla P + Q \nabla \frac{\partial H}{\partial Q} =$$

$$(-1)^{r-1}[z_r, (z_{r+1}.(\dots .(z_\ell.Q)\dots) \nabla (z_{r-1}.(\dots .(z_1.P)\dots)] - \tag{8.105}$$

$$- (z_1.(\dots .(z_\ell.Q)\dots) \nabla P + \tag{8.106}$$

$$(-1)^\ell Q \nabla (z_\ell.(\dots .(z_1.P)\dots). \tag{8.107}$$

By Lemma 7.145, (8.105) equals to the sum of two telescopic expressions the first of which cancels out (8.106) and the scond one cancelling out (8.107). Thus, the case (8.100.2) is verified, and the induction step (8.99) has been completed. Theorem 8.71 is proved. ∎

The last task of this section is to justify (7.162).

8.108. Lemma. If $H \in \mathcal{N}$ then

$$\frac{\partial H}{\partial g^a} \in \mathcal{F}^\vee \underset{\mathcal{A}_0}{\otimes} \mathcal{N}, \tag{8.109a}$$

$$\frac{\partial H}{\partial P^{\mathrm{k}_s}} \in \mathcal{A}^{\mathrm{k}+} \underset{\mathcal{A}_0}{\otimes} \mathcal{N}, \tag{8.109b}$$

$$\frac{\partial H}{\partial Q^{\mathrm{k}_s}} \in \mathcal{A}^{\mathrm{k}-} \underset{\mathcal{A}_0}{\otimes} \mathcal{N}. \tag{8.109c}$$

Proof. Since $\frac{\partial}{\partial g^a}$, etc., are all derivations, it is enough to consider H's of the form (8.78). Further, due to formulae (8.83) in the Shapovalov trick, it is enough to consider $H \in$ Plin. Let us show that for $H \in$ Plin

(8.110a) $\frac{\partial H}{\partial g^a} \in \mathcal{F}^\vee$,

(8.110b) $\frac{\partial H}{\partial P^{k_s}} \in \mathcal{A}^{k+}$,

(8.110c) $\frac{\partial H}{\partial Q^{k_s}} \in \mathcal{A}^{k-}$.

This will imply (8.109). If H is of the form (8.86) then (8.110) is obviously satisfied. Any other H from Plin can be taken to be of the form (8.94). Let us show that for such H, one has

(8.111a) $\frac{\partial H}{\partial g^c} \in \mathcal{F}^\vee$, g^c in a,

(8.111b) $\frac{\partial H}{\partial P^{k_s}} \in \mathcal{A}^{k+}$, P^{k_s} in a, $\qquad H = (a, b) \in$ Plin,

(8.111c) $\frac{\partial H}{\partial Q^{k_s}} \in \mathcal{A}^{k-}$, Q^{k_s} in a,

where "g^c in a", etc., means that g^c can be found in the expression of a through g's, P's, and Q's. Since $(a, b) = (b, a)$, (8.111) will, in turn, imply (8.110). We use induction on $rk\, a$. For $rk\, a = 1$, $a = g^d$ for some d (disregarding $\mathcal{A}_0$), and (8.111) is satisfied. For $rk\, a \geq 2$, either

(8.112.1) $a = [\varphi, \psi]$,

or

(8.112.2) $a = \varphi \nabla \psi$.

Since for (8.112.1)

(8.113) $([\varphi, \psi], b) = (\varphi, [\psi, b])$

and $rk\,\varphi < rk\,([\varphi,\psi])$, the case $a = [\varphi,\psi]$ (corresponding to the first case in (8.63)) is amenable to the induction on $rk\,a$.

For the second case (8.112.2), we have

$$(8.114)\quad (\varphi \bigtriangledown \psi, b) = <\psi, b.\varphi> = - <b.\psi, \varphi>\,.$$

If $rk\,a = 2$, so that $\varphi \bigtriangledown \psi = Q^{\mathrm{k}_s} \bigtriangledown P^{\mathrm{k}_r}$, then (8.114) yields (8.111). If $rk\,a > 2$ then from (8.64) we see that

$$(8.115)\quad \varphi = x_1.(\ldots\ .(x_p.Q^{\ell s})\ldots),$$
$$\psi = y_1.(\ldots\ .(y_q.P^{\ell\epsilon})\ldots),\qquad \text{some } x\text{'s, } y\text{'s in } \mathcal{F}^{\vee},$$

if we agree to ignore x's or y's when either are absent. So,

$$(8.116)\quad H = (\varphi \bigtriangledown \psi, b) =$$
$$((x_1.(\ldots\ .(x_p.Q^{\ell s})\ldots) \bigtriangledown (y_1.(\ldots\ .(y_q.P^{\ell\epsilon})\ldots), b) =$$
$$= < y_1.(\ldots\ .(y_q.P^{\ell\epsilon})\ldots), b.(x_1.(\ldots\ .(x_p.Q^{\ell s})\ldots) >=$$
$$(8.117a)\quad (-1)^q < P^{\ell\epsilon}, y_q.(\ldots\ .(y_1.(b.(x_1.(\ldots\ .(x_p.Q^{\ell s})\ldots) >=$$
$$(8.117b)\quad (-1)^{p+1} < x_p.(\ldots\ .(x_1.(b.(y_1.(\ldots\ .(y_q.P^{\ell\epsilon})\ldots), Q^{\ell s} >=$$
$$(8.117c)\quad (-1)^r(y_r, (y_{r-1}.(\ldots\ .(y_1.(b.(x_1.(\ldots\ .(x_p.Q^{\ell s})\ldots)\bigtriangledown$$
$$(y_{r+1}.(\ldots\ .(y_q.P^{\ell\epsilon})\ldots)) =$$
$$(8.117d)\quad (-1)^r(x_r, (x_{r+1}.(\ldots\ .(x_p.Q^{\ell s})\ldots)\bigtriangledown$$
$$(x_{r-1}.(\ldots\ .(x_1.(b.(y_1(\ldots\ .(y_q.P^{\ell\epsilon})\ldots)).$$

We see from (8.117a) that $\dfrac{\partial H}{\partial P^{\ell\epsilon}} \in \mathcal{A}^{\ell +}$ and from (8.117b) that $\dfrac{\partial H}{\partial Q^{\ell s}} \in \mathcal{A}^{\ell -}$.

Since $rk\, y_r$, $rk\, x_r < rk\, \varphi + rk\, \psi - 1 = rk\, a - 1$, (8.117c) and (8.117d) take care of all the variables entering y's and x's respectively. Thus, all the variables entering into $a = \varphi \bigtriangledown \psi$ are covered. ∎

8.118. Corollary. Formulae (7.162b) are justified by $\{(\text{Lemma } 8.108) \underset{\mathcal{A}_0}{\otimes} \widetilde{K}\}$ via specialization (8.74).

§9. Bi-SuperHamiltonian Systems

In this section we develop the theory of bi-superHamiltonian systems, thus providing the foundation for the last portion of the integrability Proof of the KdV systems constructed in §7. As an additional application, two examples, of the Harry Dym equation and its super generalization, are discussed.

We use notation of §§2,3. Let $P, Q: C \to C$ be two operators, $C = K[q_i^{(g|\sigma)}]$ (see Definition 2.33).

9.1. Lemma.

$$(9.2)\quad (PQ)^{\dagger} = Q^{\dagger} P^{\dagger} (-1)^{p(P)p(Q)}.$$

Proof. For any $u, v \in C$, we have, by repeatedly using (2.37):

$$[(PQ)^{\dagger}(v)]u(-1)^{p(u)p(v)} \sim [(PQ)(u)]v = [P(Q(u))]v \sim$$

(9.3)

$$\sim [P^{\dagger}(v)][Q(u)](-1)^{p(v)[p(Q)+p(u)]} = [Q(u)][P^{\dagger}(v)](-1)^{p(P)[p(Q)+p(u)]} \sim$$

$$\sim [Q^{\dagger}(P^{\dagger}(v))](u)(-1)^{p(P)p(Q)+p(u)p(v)}.$$

Comparing the end terms in (9.3) and using Lemma 2.38, we arrive at (9.2). ∎

9.4. Lemma. Let $P, Q: C^N \to C^N$, $N = |I|$, be two even operators. Then

$$(9.5)\quad (PQ)^{s\dagger} = Q^{s\dagger} E P^{s\dagger},$$

where

$$(9.6)\quad E = \mathbf{1}^{s\dagger} = \text{diag}\,(\mathbf{1}\ ,\mathbf{-1}):\quad E_{ij} = (-1)^{p(i)}\delta_{ij}\,.$$

Proof. By (2.43), we have

$\left((PQ)^{s\dagger}\right)_{ij} = (-1)^{p(i)p(j)}\,[(PQ)_{ji}]^{\dagger} = (-1)^{p(i)p(j)}\left(P_{jk}Q_{ki}\right)^{\dagger}$ [by (9.2) and since P, Q are even] $= (-1)^{p(i)\,p(j)+[p(j)+p(\mathrm{k})][p(\mathrm{k})+p(i)]} Q^{\dagger}_{\mathrm{k}i} \, P^{\dagger}_{j\mathrm{k}}$

[by (2.43)] $= (Q^{s\dagger})_{i\mathrm{k}}(-1)^{p(\mathrm{k})}(P^{s\dagger})_{\mathrm{k}j}$ [by(9.6)] $= (Q^{s\dagger} E P^{s\dagger})_{ij}$. ∎

9.7. Definition. Two superHamiltonian matrices $B^I$and B^{II} over C form a *superHamiltonian pair* if

(9.8) $\alpha B^I + \beta B^{II}$ is Hamiltonian , $\forall \alpha, \beta \in (K')_c$, for any extension $K' \supset K$.

Now we can state the main result of this section.

9.9. Theorem. Let B^I , B^{II} be a superHamiltonian pair. Let $X, Y, Z \in C^N$ be even finite vectors satisfying

(9.10a) $B^I(Y) = B^{II}(X)$,

(9.10b) $B^I(Z) = B^{II}(Y)$.

Suppose B^I is nondegenerate in the sense that if $\mathcal{M}$ is an even $N \times N$ matrix operator over C then

(9.11) $B^I \mathcal{M} B^I = 0$ implies $\mathcal{M} = 0$.

Conclusion: if $X, Y \in Im\delta$ then so is Z, where $\delta\colon C \to \Omega^1_0(C) \approx C^N$ is the Euler-Lagrange operator (see §2.).

Proof. Extend K into $K' = K[\theta]/(\theta^2+1)$, $p(\theta) = 0 \in \mathbf{Z}_2$, $\theta \in (K')_c$. In other words, "$\theta = \sqrt{-1}$", but the letter i is already occupied marking various q's. Extend correspondingly C into C', etc. Set

(9.12) $$E^{+} = \operatorname{diag}(\mathbf{1}, \theta\mathbf{1}) : \quad (E^{+})_{ij} = \begin{cases} \delta_{ij}, & p(i) = 0, \\ \delta_{ij}\theta, & p(i) = 1, \end{cases}$$

(9.13) $E^- = E\,E^+ = E^+E\,,$

so that

(9.14) $(E^+)^2 = (E^-)^2 = E\,,$

(9.15) $E\,E^- = E^-E = E^+\,,\quad E^+\,E^- = E^-\,E^+ = \mathbf{1}\,,$

(9.16) $(E^+)^{s\dagger} = E^-\,, (E^-)^{s\dagger} = E^+\,.$

For an even $N \times N$ matrix operator P, define

(9.17) $\widetilde{P} = E^-P\,E^-\,.$

9.18. Lemma. $\widetilde{P}$ is superskewsymmetric iff P is.

Proof. By (9.5),

$$(\widetilde{P})^{s\dagger} = (E^-)^{s\dagger}E\,P^{s\dagger}E(E^-)^{s\dagger} \text{ [by (9.16), (9.13)] } =$$
$$= E^-P^{s\dagger}E^- = \widetilde{P^{s\dagger}}\,.$$

Thus, if $P^{s\dagger} = -P$ then $\widetilde{P}^{s\dagger} = -\widetilde{P}$. Conversely, if $\widetilde{P}^{s\dagger} = -\widetilde{P}$ then $\widetilde{P^{s\dagger}} = -\widetilde{P}$ so that $P^{s\dagger} = -P\,.$ ∎

For an even vector $X \in C'^N$, define

(9.19) $\widetilde{D}(X) = E^+D^0(X)E^-\,.$

9.20. Lemma. $\widetilde{D}(X)$ is supersymmetric iff $D(X)$ is.

Proof.By(9.5), (9.16), (9.19),

$$[\widetilde{D}(X)]^{s\dagger} = E^+E\,[D^0(X)]^{s\dagger}\,E\,E^-\,,$$

so that

$$E^+ D^0(X) E^- = \widetilde{D}(X) = [\widetilde{D}(X)]^{s\dagger} = E^+ E[D^0(X)]^{s\dagger} E E^-$$

iff $D^0(X) = E[D^0(X)]^{s\dagger} E$, or, equivalently, iff

$$E D^0(X) E = [D^0(X)]^{s\dagger} . \tag{9.21}$$

Now, by (9.6), (2.26), and (2.21),

$$(ED^0(X)E)_{ij} =$$

$$= (-1)^{p(i)}(-1)^{p(j)[p(i)+1]} D_j(X_i)(-1)^{p(j)} = (-1)^{p(i)[p(j)+1]} D_j(X_i) =$$

$$(-1)^{p(i)[p(j)+1]} [D(X)]_{ij} , \tag{9.22l}$$

while

$$\left([D^0(X)]^{s\dagger}\right)_{ij} = (-1)^{p(i)p(j)} \left([D^0(X)]_{ji}\right)^{\dagger} =$$

$$= (-1)^{p(i)p(j)}(-1)^{p(i)[p(j)+1]}[D_i(X_j)]^{\dagger} =$$

$$= (-1)^{p(i)[p(j)+1]}(-1)^{p(i)p(j)} \left([D(X)]_{ji}\right)^{\dagger} =$$

$$(-1)^{p(i)[p(j)+1]} \left([D(X)]^{s\dagger}\right)_{ij} . \tag{9.22r}$$

Comparing (9.22l) with (9.22r)we see that (9.21)holds true iff

$D(X) = [D(X)]^{s\dagger}$, as required. ∎

9.23. Lemma. $\widetilde{P}$ is nondegenerate iff P is.

Proof. $\widetilde{P} \mathcal{M} \widetilde{P} = E^-[P(E^- \mathcal{M} E^-)P]E^-$, and E^- is invertible. ∎

Applying D^0 to each of the two equalities (9.10), we obtain

$$B^I D^0(Y) + [D^0, B^I](Y) = B^{II} D^0(X) + [D^0, B^{II}](X) , \tag{9.24a}$$

$$B^I D^0(Z) + [D^0, B^I](Z) = B^{II} D^0(Y) + [D^0, B^{II}](Y) . \tag{9.24b}$$

Multiplying each of the equations (9.24) by E^- from the right and left, we get

$$\widetilde{B}^I \widetilde{D}(Y) + E^-[D^0, B^I](Y)E^- = \widetilde{B}^{II} \widetilde{D}(X) + E^-[D^0, B^{II}](X)E^- , \tag{9.25a}$$

(9.25b) $\widetilde{B}^I\widetilde{D}(Z) + E^-[D^0, B^I](Z)E^- = \widetilde{B}^{II}\widetilde{D}(Y) + E^-[D^0, B^{II}](Y)E^-$.

Multiply from the right: (9.25a) by $E\widetilde{B}^{II}$, and (9.25b) by $E\widetilde{B}^I$, resulting in

(9.26a) $\widetilde{B}^I\widetilde{D}(Y)E\widetilde{B}^{II} + E^-[D^0, B^I](Y)E^+\widetilde{B}^{II} =$

$\widetilde{B}^{II}\widetilde{D}(X)E\widetilde{B}^{II} + E^-[D^0, B^{II}](X)E^+\widetilde{B}^{II}$,

(9.26b) $\widetilde{B}^I\widetilde{D}(Z)E\widetilde{B}^I + E^-[D^0, B^I](Z)E^+\widetilde{B}^I =$

$\widetilde{B}^{II}\widetilde{D}(Y)E\widetilde{B}^I + E^-[D^0, B^{II}](Y)E^+\widetilde{B}^I$.

Subtracting (9.26a) from (9.26b), we earn

(9.27a) $\widetilde{B}^I\widetilde{D}(Z)E\widetilde{B}^I -$

(9.27b) $-[\widetilde{B}^I\widetilde{D}(Y)E\widetilde{B}^{II} + \widetilde{B}^{II}\widetilde{D}(Y)E\widetilde{B}^I] +$

(9.27c) $+\widetilde{B}^{II}\widetilde{D}(X)E\widetilde{B}^{II} =$

(9.28)
$$E^-\{[D^0, B^{II}](Y)E^+\widetilde{B}^I - [D^0, B^{II}](X)E^+\widetilde{B}^{II} + \\ +[D^0, B^I](Y)E^+\widetilde{B}^{II} - [D^0, B^I](Z)E^+\widetilde{B}^I\}.$$

Our plan now is this: we are going to show that all the expressions (9.27b,c) and (9.28) are supersymmetric. This will imply that (9.27a) is supersymmetric as well, and this will be close enough to home.

9.29. Lemma. If $B, \mathcal{M}$ are even $N \times N$ matrix operators and B is superskewsymmetric then

(9.30) $(B\mathcal{M}EB)^{s\dagger} = B\mathcal{M}^{s\dagger}EB$.

Proof. By (9.5),

$(B\mathcal{M}EB)^{s\dagger} = B^{s\dagger}EE^{s\dagger}E\mathcal{M}^{s\dagger}EB^{s\dagger} = B\mathcal{M}^{s\dagger}EB.$ ∎

9.31. Corollary. (9.27c) is supersymmetric.

Proof. B^{II} is superskewsymmetric, hence $\widetilde{B}^{II}$ is superskewsymmetric by Lemma 9.18. Further, $X \in Im\delta$ so that $D(X)$ is supersymmetric by (2.66); hence, $\widetilde{D}(X)$ is also supersymmetric thanks to Lemma 9.20. Now apply (9.30) with $B = \widetilde{B}^{II}$ and $\mathcal{M} = \widetilde{D}(X)$. ∎

9.32. Lemma. If $B_1, B_2, \mathcal{M}$ are even $N \times N$ matrix operators and B_1, B_2 are superskewsymmetric then

$$(9.33)\quad (B_1 \mathcal{M} E B_2)^{s\dagger} = B_2 \mathcal{M}^{s\dagger} E B_1 .$$

Proof. By (9.5),

$$(B_1 \mathcal{M} E B_2)^{s\dagger} = B_2^{s\dagger} E E^{s\dagger} E \mathcal{M}^{s\dagger} E B_1^{s\dagger} = B_2 \mathcal{M}^{s\dagger} E B_1.$$ ∎

9.34. Corollary. (9.27b) is supersymmetric.

Proof is analogous to that of Corollary 9.31. ∎

9.35. Lemma. (9.28) is supersymmetric.

Granted Lemma 9.35, we find that (9.27a) is supersymmetric. Thus,

$$0 = \widetilde{B}^I \widetilde{D}(Z) E \widetilde{B}^I - (\widetilde{B}^I \widetilde{D}(Z) E \widetilde{B}^I)^{s\dagger} \text{ [by (9.30) and Lemma 9.18] } =$$
$$= \widetilde{B}^I (\{\widetilde{D}(Z) - [\widetilde{D}(Z)]^{s\dagger}\} E) \widetilde{B}^I .$$

Therefore, by Lemma 9.23, $(\widetilde{D}(Z) - [\widetilde{D}(Z)]^{s\dagger})E = 0$ and since E is invertible, $\widetilde{D}(Z) = [\widetilde{D}(Z)]^{s\dagger}$. By Lemma 9.20 it follows that $D(Z) = [D(Z)]^{s\dagger}$. This in turn implies, by Theorem 2.67, that $Z \in Im\delta$. Theorem 9.9 is proved. ∎

Proof of Lemma 9.35. Denote the expression (9.28) by V. To compute $V^{s\dagger}$ notice that if B is superskewsymmetric then

$$(E^- \mathcal{M} E^+ \widetilde{B})^{s\dagger} = -\widetilde{B} E E^- E \mathcal{M}^{s\dagger} E E^+ = -\widetilde{B} E^- \mathcal{M}^{s\dagger} E^- =$$

$$= -E^{-}BE\mathcal{M}^{s\dagger}E^{-}.$$

Hence,

$$(9.36)\quad V^{s\dagger} = E^{-}U_1E^{-},$$

where

$$(9.37)\quad \begin{aligned} U_1 = -B^{I}E([D^0, B^{II}](Y))^{s\dagger} + B^{II}E([D^0, B^{II}](X))^{s\dagger} - \\ - B^{II}E([D^0, B^{I}](Y))^{s\dagger} + B^{I}E([D^0, B^{I}](Z))^{s\dagger}. \end{aligned}$$

Write (9.28) in the form $V = E^{-}U_2\,E^{-}$, where

$$(9.38)\quad \begin{aligned} U_2 = [D^0, B^{II}](Y)B^{I} - [D^0, B^{II}](X)B^{II} + \\ + [D^0, B^{I}](Y)B^{II} - [D^0, B^{I}](Z)B^{I}. \end{aligned}$$

To show that $V^{s\dagger} = V$, we have to demonstrate that $U_1 = U_2$.

By Theorem 3.52, a matrix B is superHamiltonian iff

$$(9.39)\quad BE([D^0, B](X))^{s\dagger}(Y) = ([D^0, B](Y))B(X) - ([D^0, B](X))B(Y),$$

for any even finite vectors X and Y, since (9.39) is nothing but (3.53) with the term $< B, R, S >$ replaced, thanks to (3.44) and (9.6), by $([D^0, B](R))^{s\dagger}(S)$. Applying (9.39) to the case $B = B^{I} + \epsilon B^{II}$ in $K' = K[\epsilon], p(\epsilon) = 0, \epsilon \in (K')_c$, and equating the resulting terms in front of ϵ^0, ϵ^2, and ϵ^1, we obtain

$$(9.40)\quad B^{I}E([D^0, B^{I}](Z))^{s\dagger}(R) = ([D^0, B^{I}](R))B^{I}(Z) - ([D^0, B^{I}](Z))B^{I}(R),$$

$$(9.41)\quad \begin{aligned} & B^{II}E([D^0, B^{II}](X))^{s\dagger}(R) \\ & = ([D^0, B^{II}](R))B^{II}(X) - ([D^0, B^{II}](X))B^{II}(R), \end{aligned}$$

$$9.42)\quad \begin{aligned} & B^{II}E([D^0, B^{I}](Y))^{s\dagger}(R) + B^{I}E([D^0, B^{II}](Y))^{s\dagger}(R) = \\ & = ([D^0, B^{I}](R))B^{II}(Y) + ([D^0, B^{II}](R))B^{I}(Y) - \\ & -([D^0, B^{I}](Y))B^{II}(R) - ([D^0, B^{II}](Y))B^{I}(R), \end{aligned}$$

for any even finite vectors X, Y, Z, R. Now, applying the operator U_1 (9.37) to arbitrary even finite vector R and using (9.40)+(9.41)−(9.42), we get $U_1(R) =$

(9.43a, b) $([D^0, B^I](R))B^I(Z) - ([D^0, B^I](Z))B^I(R)+$

(9.44a, b) $([D^0, B^{II}](R))B^{II}(X) - ([D^0, B^{II}](X))B^{II}(R)-$

(9.45a, b) $-([D^0, B^I](R))B^{II}(Y) - ([D^0, B^{II}](R))B^I(Y)+$

(9.46a, b) $([D^0, B^I](Y))B^{II}(R) + ([D^0, B^{II}](Y))B^I(R)$.

On the other hand, from (9.38) we obtain $U_2(R) =$

(9.47a, b) $([D^0, B^{II}](Y))B^I(R) - ([D^0, B^{II}](X))B^{II}(R)+$

(9.48a, b) $([D^0, B^I](Y))B^{II}(R) - ([D^0, B^I](Z))B^I(R)$.

Now, (9.43a) + (9.45a) = 0 by (9.10b), and (9.44a) + (9.45b) = 0 by (9.10a). The remaining terms in $U_1(R)$ and $U_2(R)$ match into the following pairs: (9.43b) = (9.48b), (9.44b) = (9.47b), (9.46a) = (9.48a), (9.46b) = (9.47a). Thus, $U_1(R) = U_2(R)$ for any finite even R. From Lemma 3.19 it follows that $U_1 = U_2$.

Thus, $V^{s\dagger} = V$. ∎

Now we can justify various claims, similar to the conclusion of Theorem 9.9, made in §7. First, in all three cases: (7.55) and (7.56); (7.87) and (7.88); (7.114) and (7.119, 7.120), - we have superHamiltonian pairs, since in all three cases our situation is of the type $B^{II} = B(\mathcal{L}) + b_\mu, B^I = b_\nu$, (see (3.111)) where $\mathcal{L}$ is a differential Lie superalgebra and μ, ν are generalized 2-cocycles on $\mathcal{L}$. Second, the matrices B^I given by: (7.55); (7.87); (7.114), - are obviously nondegenerate in the sense of Theorem 9.9. And that's it.

9.49. Lemma. Let B_1, B_2 be two superHamiltonian matrices over C, and let

$H_1, H_2, \ldots$ be even elements in C satisfying

$$(9.50) \quad B_1(X_{r+1}) = B_2(X_r), \quad 1 \leq r \leq M, \quad M \in \mathbf{N} \cup \{\infty\},$$

where $X_r = \delta(H_r) \in C^N$. Denote by $\{,\}_j, \quad j = 1, 2$, the Poisson bracket in C corresponding to the superHamiltonian matrix B_j. Then

$$(9.51) \quad \{H_\alpha, H_\beta\}_j \sim 0, \quad \forall \alpha, \beta \in \{1, \ldots, M+1\}, \ \forall j \in \{1, 2\}.$$

<u>Proof</u>. Indeed, (9.51) is true for $\alpha = \beta$ since B_j is superskewsymmetric. If $\alpha \neq \beta$ then $\{H_\alpha, H_\beta\}_j \sim -\{H_\beta, H_\alpha\}_j$, so that we can consider only the case $\alpha < \beta$, say. We have, using $\{H_\alpha, H_\beta\}_j \sim B_j(X_\alpha)^t(X_\beta)$:

$\{H_\alpha, H_{\alpha+r}\}_1 \sim B_1(X_\alpha)^t X_{\alpha+r} \sim -B_1(X_{\alpha+r})^t X_\alpha = -B_2(X_{\alpha+r-1})^t X_\alpha \sim$

$$(9.52) \quad \sim B_2(X_\alpha)^t X_{\alpha+r-1} \sim \{H_\alpha, H_{\alpha+r-1}\}_2 \, ,$$

$$(9.53) \quad \{H_\alpha, H_{\alpha+r}\}_1 \sim B_2(X_\alpha)^t X_{\alpha+r-1} = B_1(X_{\alpha+1})^t X_{\alpha+r-1} \sim$$

$\{H_{\alpha+1}, H_{\alpha+r-1}\}_1$.

From (9.53) we find that

$\{H_\alpha, H_{\alpha+2r}\}_1 \sim \{H_{\alpha+r}, H_{\alpha+r}\}_1 \sim 0$,

$\{H_\alpha, H_{\alpha+2r-1}\}_1 \sim \{H_{\alpha+r-1}, H_{\alpha+r}\}_1$ [by (9.52)] $\sim$

$\{H_{\alpha+r-1}, H_{\alpha+r-1}\}_2 \sim 0$,

so that, indeed, $\{H_\alpha, H_\beta\}_1 \sim 0, \forall \alpha, \beta$. Also, from (9.52) we obtain

$\{H_\alpha, H_\beta\}_2 \sim \{H_\alpha, H_{\beta+1}\}_1 \sim 0$ for $\beta \leq M$.

Finally,

$\{H_\alpha, H_{M+1}\}_2 \sim B_2(X_\alpha)^t X_{M+1} = B_1(X_{\alpha+1})^t X_{M+1} \sim \{H_{\alpha+1}, H_{M+1}\}_1 \sim 0$. ∎

9.54. Remark. In Lemma 9.49: B_1 and B_2 do *not* have to form a superHamiltonian pair; and the favorite boundary condition of §7, $B_1(\delta H_1) = 0$, is *not* required.

We conclude this section with two instructive examples of bi-superHamiltonian systems, of a kind different from the ones considered in §7, namely, of the type

$$(9.55) \quad B^I = B(\mathcal{L}), \quad B^{II} = b_\nu, \quad H_0 \in Ker B^I,$$

i.e., when H_0 is a "Casimir element" for a Lie superalgebra $\mathcal{L}$. (In practice, B^{II} is nondegenerate in the sense that $Ker B^{II}$ is not a K'-submodule in C', for any $K' \supset K$.)

We start with the following simple case first. Let K be a commutative differential algebra and let $\mathcal{L} = D = D(K) = \{0\}_{\text{aff}}$ (see (7.32)). Then $B^I = B(D)$ is the upper left corner of the matrix (7.50):

$$(9.56) \quad B^I = u\partial + \partial u\,.$$

Consider the 2-cocycle (7.46) on $D(K)$, so that

$$(9.57) \quad B^{II} = \partial^3\,.$$

Pick out

$$(9.58) \quad H_0 = 2u^{1/2}\,.$$

Then

$$(9.59) \quad B^I \delta(H_0) = (u\partial + \partial u)(u^{-1/2}) = 2u^{1/2}\partial u^{1/2}(u^{-1/2}) = 0\,,$$

and we are in the situation (9.55). The equation

$$(9.60) \quad u_t = B^{II}\delta(H_0) = \partial^3(u^{-1/2})$$

is called the (Harry) Dym equation. Let us show that it is the first member of an infinite bi-Hamiltonian hierarchy. Write

(9.61) $B^I(X_{r+1}) = B^{II}(X_r), \quad r \in \mathbf{Z}_+ ,$

with $X_0 = u^{-1/2}$. We want to show that (9.61) can be solved step by step. Let us use induction on r. To avoid the labor of checking the case $r = 0$ in (9.61), extend r into $\{-1\} \cup \mathbf{Z}_+$ and set

(9.62) $H_{-1} = 0$

to satisfy (9.61) for $r = -1$. Now rewrite (9.61) in the long hand as

$(u\partial + \partial u)(X_{r+1}) = \partial^3(X_r),$

or, since $u\partial + \partial u = 2u^{1/2}\partial u^{1/2}$, as

(9.63) $2\partial(u^{1/2}X_{r+1}) = u^{-1/2}\partial^3(X_r)\,.$

To solve (9.63) we show that, if $X_{-1}, X_0, ..., X_r$ are constructed, then $u^{-1/2}\partial^3(X_r) \sim 0$. By Theorem 9.9, there exist $H_{-1}, ..., H_r$ such that $X_i = \delta(H_i), -1 \leq i \leq r$.Hence,

(9.64) $u^{-1/2}\partial^3(X_r) \sim -\partial^3(u^{-1/2})X_r = -\{H_0, H_r\}_2$ [by Lemma 9.49] $\sim 0,$

as desired, and we are done.

9.65. Remark. We have operated above with expressions involving $u^{1/2}$ and $u^{-1/2}$, and they are not of polynomial type. To justify the handling, one should make minor but numerous adjustments in all the machinery of the calculus of variations, superHamiltonian formalism, etc., to incorporate rational and algebraic expressions involving $q_i^{(e|0)}$'s ; alternatively, by introducing the variable $v = u^{1/2}$, one could avoid algebraic irrationals since in the v-language (9.56)-(9.58) become

(9.66) $$B^I = \frac{1}{2}\partial, \quad B^{II} = \frac{1}{4}v^{-1}\partial^3 v^{-1}, H_0 = 2v,$$

with only rationals involved. We shall willfully overlook this problem trusting the reader to fix up his misgiving should he have any.

For our second example, let K be a differential commutative superalgebra, and let $\mathcal{L} = D^s = D^s(K) = \{0\}^s_{\text{aff}}$ (see (7.33)). Then $B^I = B^I(D^s)$ is the following submatrix of the matrix (7.50)

(9.66) $$B^I = \begin{pmatrix} u\partial + \partial u & \varphi\partial + \frac{1}{2}\partial\varphi \\ \partial\varphi + \frac{1}{2}\varphi\partial & 2u \end{pmatrix}$$

Taking the 2-cocycle (7.37) on D^s, we obtain (see (7.51)) :

(9.67) $$B^{II} = \text{diag}\,(\partial^3, 4\partial^2).$$

9.68. Lemma. Set

(9.69) $$H_0 = 2u^{1/2} - \frac{1}{4}u^{-3/2}\varphi\varphi^{(1)}.$$

Then $H_0 \in Ker B^I$.

Proof. We have

(9.70a) $$\frac{\delta H_0}{\delta u} = u^{-1/2} + \frac{3}{8}u^{-5/2}\varphi\varphi^{(1)},$$

and also $\dfrac{\delta H_0}{\delta\varphi} = -\dfrac{1}{4}u^{-3/2}\varphi^{(1)} - \partial\left(\dfrac{1}{4}u^{-3/2}\varphi\right)$:

(9.70b) $$\frac{\delta H_0}{\delta\varphi} = -\frac{1}{2}u^{-3/2}\varphi^{(1)} + \frac{3}{8}\varphi u^{-5/2}u^{(1)}.$$

Now,

$$(u\partial + \partial u)\left(\frac{\delta H_0}{\delta u}\right) + \left(\varphi\partial + \frac{1}{2}\partial\varphi\right)\left(\frac{\delta H_0}{\delta\varphi}\right) =$$

$$= (u\partial + \partial u)\left(u^{-1/2} + \frac{3}{8}u^{-5/2}\varphi\varphi^{(1)}\right) +$$

$$\left(\varphi\,\partial + \frac{1}{2}\partial\varphi\right)\left[-\frac{1}{2}u^{-3/2}\varphi^{(1)} + \frac{3}{8}\varphi u^{5/2}u^{(1)}\right] =$$

$$= (2\partial u - u^{(1)})\left(\frac{3}{8}u^{-5/2}\varphi\varphi^{(1)}\right) +$$

$$\left(\frac{3}{2}\partial\varphi - \varphi^{(1)}\right)\left[-\frac{1}{2}u^{-3/2}\varphi^{(1)} + \frac{3}{8}\varphi u^{-5/2}u^{(1)}\right] =$$

$$= \frac{3}{4}\partial(u^{-3/2}\varphi\varphi^{1}) - \frac{3}{8}u^{(1)}u^{-5/2}\varphi\varphi^{(1)} - \frac{3}{4}\partial(\varphi u^{-3/2}\varphi^{(1)}) - \frac{3}{8}\varphi^{(1)}\varphi u^{-5/2}u^{(1)} = 0,$$

and

$$(\partial\varphi + \frac{1}{2}\varphi\partial)(\frac{\delta H_0}{\delta u}) + 2u\frac{\delta H_0}{\delta\varphi} =$$

$$= (\frac{3}{2}\partial\varphi - \frac{1}{2}\varphi^{(1)})(u^{-1/2} + \frac{3}{8}u^{-5/2}\varphi\varphi^{(1)}) + 2u(-\frac{1}{2}u^{-3/2}\varphi^{(1)} + \frac{3}{8}\varphi u^{-5/2}u^{(1)}) =$$

$$= \frac{3}{2}\partial(\varphi u^{-1/2}) - \frac{1}{2}u^{-1/2}\varphi^{(1)} - u^{-1/2}\varphi^{(1)} + \frac{3}{4}\varphi u^{-3/2}u^{(1)} = 0. \qquad \blacksquare$$

We now define the system $B^{II}\delta(H_0)$:

$$(9.71) \qquad \begin{cases} u_t = \partial^3\left(u^{-1/2} + \frac{3}{8}u^{-5/2}\varphi\varphi^{(1)}\right), \\ \varphi_t = 4\partial^2\left(-\frac{1}{2}u^{-3/2}\varphi^{(1)} + \frac{3}{8}\varphi u^{-5/2}u^{(1)}\right), \end{cases}$$

which can be thought of as a supergeneralization of the Dym equation (9.60) to which it reduces when φ vanishes in (9.71).

We are going to show that the supersystem (9.71) belongs to an infinite bi-superHamiltonian hierarchy.

Set

$$(9.72)\quad X_r = \begin{pmatrix} Y_r \\ \alpha_r \end{pmatrix}, \quad r \in \{-1\} \cup \mathbf{Z}_+ \,, \; p(Y_r) = 0, \; p(\alpha_r) = 1,$$

with

$$(9.73)\quad X_{-1} = \begin{pmatrix} 0 \\ 0 \end{pmatrix} \quad , \quad X_0 = \begin{pmatrix} \delta H_0/\delta u \\ \delta H_0/\delta \varphi \end{pmatrix} ,$$

where $\delta(H_0)$ is given by (9.70). To solve the recursive relation $B^I \delta(H_{r+1}) = = B^{II}\delta(H_r)$ in the form

$$(9.74)\qquad B^I(X_{r+1}) = B^{II}(X_r) \,, \quad r \geq -1 \,,$$

with $X_r = \delta(H_r)$, we use induction on r. The case $r = -1$ is satisfied since $B^I\delta(H_0) = 0$. Let us rewrite the equation (9.74) in the long hand as

$$(9.75a)\quad (u\partial + \partial u)(Y_{r+1}) + \left(\varphi\,\partial + \frac{1}{2}\partial\varphi\right)(\alpha_{r+1}) = \partial^3(Y_r) \,,$$

$$(9.75b)\quad \left(\partial\varphi + \frac{1}{2}\varphi\partial\right)(Y_{r+1}) + 2u\alpha_{r+1} \;\; = 4\partial^2(\alpha_r) \,.$$

Solving (9.75b) in favour of α_{r+1} and substituting the resulting expression into (9.75a), we obtain

$$(9.76)\quad \begin{aligned} &\left[u\partial + \partial u + \left(\varphi\partial + \frac{1}{2}\partial\varphi\right)\left(-\frac{1}{2u}\right)\left(\partial\varphi + \frac{1}{2}\varphi\partial\right)\right](Y_{r+1}) = \\ &= \partial^3(Y_r) - \left(\varphi\partial + \frac{1}{2}\partial\varphi\right)\frac{4}{2u}\partial^2(\alpha_r) \,. \end{aligned}$$

Now,

$$\begin{aligned} &\left(\varphi\partial + \frac{1}{2}\partial\varphi\right)\frac{1}{u}\left(\partial\varphi + \frac{1}{2}\varphi\partial\right) = \left(\frac{3}{2}\partial\varphi - \varphi^{(1)}\right)u^{-1}\left(\frac{3}{2}\varphi\partial + \varphi^{(1)}\right) = \\ &= \frac{3}{2}\partial\varphi u^{-1}\varphi^{(1)} - \varphi^{(1)}u^{-1}\frac{3}{2}\varphi\partial = \frac{3}{2}(\varphi\varphi^{(1)}u^{-1}\partial + \partial\varphi\varphi^{(1)}u^{-1}) \,. \end{aligned}$$

Hence,

$$u\partial + \partial u + \left(\varphi\partial + \frac{1}{2}\partial\varphi\right)\left(-\frac{1}{2u}\right)\left(\partial\varphi + \frac{1}{2}\varphi\partial\right) =$$
$$= \left(u - \frac{3}{4}u^{-1}\varphi\varphi^{(1)}\right)\partial + \partial\left(u - \frac{3}{4}u^{-1}\varphi\varphi^{(1)}\right) =$$
$$= 2\left(u - \frac{3}{4}u^{-1}\varphi\varphi^{(1)}\right)^{1/2}\partial\left(u - \frac{3}{4}u^{-1}\varphi\varphi^{(1)}\right)^{1/2} =$$
$$= 2u^{1/2}\left(1 - \frac{3}{8}u^{-2}\varphi\varphi^{(1)}\right)\partial u^{1/2}\left(1 - \frac{3}{8}u^{-2}\varphi\varphi^{(1)}\right),$$

where in the last equality of the chain we used the fact that $\varphi\varphi^{(1)}$ is nilpotent : $(\varphi\varphi^{(1)})^2 = 0$. Thus, (9.76) can be rewritten in the form

$$(9.77)\quad 2\partial[u^{1/2}\left(1 - \frac{3}{8}u^{-2}\varphi\varphi^{(1)}\right)Y_{r+1}] =$$
$$(9.78)\quad u^{-1/2}\left(1 - \frac{3}{8}u^{-2}\varphi\varphi^{(1)}\right)^{-1}[\partial^3(Y_r) - \left(\varphi\partial + \frac{1}{2}\partial\varphi\right)\frac{2}{u}\partial^2(\alpha_r)].$$

Thus, to complete the induction step we have to show that the expression (9.78) belongs to $Im\partial$ provided we have already found $X_{-1}, ..., X_r$ satisfying (9.74). By Theorem 9.9, there exist $H_{-1}, ..., H_r$ such that $X_i = \delta(H_i)$. By Lemma 9.49,

$$0 \sim \{H_r, H_0\}_2 \sim \partial^3(Y_r)\frac{\delta H_0}{\delta u} + 4\partial^2(\alpha_r)\frac{\delta H_0}{\delta\varphi}\quad [\text{by } (9.70)] =$$
$$(9.79a)\quad \partial^3(Y_r)u^{-1/2}\left(1 + \frac{3}{8}u^{-5/2}\varphi\varphi^{(1)}\right) +$$
$$(9.79b)\quad 4\partial^2(\alpha_r)u^{-3/2}\frac{1}{2}\left(-\varphi^{(1)} + \frac{3}{4}\varphi u^{-1}u^{(1)}\right),$$

and we see that the first (Y_r) term in (9.78) equals to (9.79a). Let us transform the second term in (9.78):

$$-u^{-1/2}\left(1 + \frac{3}{8}u^{-2}\varphi\varphi^{(1)}\right)\left(\varphi\partial + \frac{1}{2}\partial\varphi\right)\frac{2}{u}\partial^2(\alpha_r) =$$

$$= -u^{-1/2}\left(\varphi\partial + \frac{1}{2}\partial\varphi\right)\frac{2}{u}\partial^2(\alpha_r) \sim$$

$$\sim [(u^{-1/2}\varphi)^{(1)} + \frac{1}{2}(u^{-1/2})^{(1)}\varphi]\frac{2}{u}\partial^2(\alpha_r)[\text{since } p(\varphi) = p(\alpha_r) = 1] =$$

$$= -\frac{2}{u}\partial^2(\alpha_r)\left[u^{-1/2}\varphi^{(1)} - \frac{3}{4}u^{-3/2}u^{(1)}\varphi\right] = 2\partial^2(\alpha_r)u^{-3/2}\left(-\varphi^{(1)} + \frac{3}{4}u^{-1}u^{(1)}\varphi\right),$$

and this is the same as (9.79b). Thus, (9.78) ~ 0. ∎

Appendix. Metrizable Lie Algebras

One of the basic ingredients of the constructions in §7 is a metrizable Lie algebra $\mathcal{G}$ with a fixed orthonormal basis (see p. 107). It was not clear whether there exist many such algebras outside the class of semisimple ones. As a matter of fact there are many indeed, as we shall see presently.

We start with an observation that for results in §§7, 8 to hold true, one does not need an orthonormal basis for $\mathcal{G}$: any (orthogonal) basis in which the matrix of the invariant metric is scalar, will do, since the structure constants t^i_{jk} of $\mathcal{G}$ in such a basis will still be cyclic-symmetric.

Now, let $\mathcal{L}$ be an arbitrary finite-dimensional Lie algebra over a commutative algebra $\mathcal{A}_0 \supset \mathbf{Q}$, and suppose $\mathcal{L}$ is a free $\mathcal{A}_0$-module (see p. 114). Choose a basis $(e_1, \ldots, e_n)$ of $\mathcal{L}$ and set

$$\mathcal{G} = \mathcal{L} \propto \mathcal{L}^*. \tag{A.1}$$

where $\mathcal{L}^*$ is the dual space to $\mathcal{L}$. Let $(\bar{e}_1, \ldots, \bar{e}_n)$ be the basis of $\mathcal{L}^*$ dual to the chosen basis of $\mathcal{L}$.

Proposition A.2. (*i*) The following is an invariant metric on $\mathcal{G}$:

$$\left(\begin{pmatrix} x_1 \\ a_1 \end{pmatrix}, \begin{pmatrix} x_2 \\ a_2 \end{pmatrix}\right) = < a_1, x_2 > + < a_2, x_1 >, \qquad x_\ell \in \mathcal{L}, a_\ell \in \mathcal{L}^*, \quad \ell = 1, 2; \tag{A.3}$$

(*ii*) The vectors $e^\epsilon_j := e_j + \epsilon \bar{e}_j, \quad j = 1, \ldots, n, \quad \epsilon = \pm 1$, form an orthogonal basis of $\mathcal{G}$, and $(e^\epsilon_j, e^\epsilon_j) = 2, \ 1 \leq j \leq n, \ \epsilon = \pm 1$.

Proof. (ii) follows from (A.3). Now,

$$([1,2],3) = (\begin{pmatrix} [x_1, x_2] \\ x_1.a_2 - x_2.a_1 \end{pmatrix}, \begin{pmatrix} x_3 \\ a_3 \end{pmatrix}) =< x_1.a_2 - x_2.a_1, x_3 > +$$

$$+ < a_3, [x_1, x_2] >= - < a_2, [x_1, x_3] > + < a_1, [x_2, x_3] > + < a_3, [x_1, x_2] >=$$

$$=< a_1, [x_2, x_3] > - < a_3, [x_2, x_1] > + < a_2, [x_3, x_1] >=$$

$$=< a_1, [x_2, x_3] > + < x_2.a_3 - x_3.a_2, x_1 >=$$

$$= (\begin{pmatrix} x_1 \\ a_1 \end{pmatrix}, \begin{pmatrix} [x_2, x_3] \\ x_2.a_3 - x_3.a_2 \end{pmatrix}) = (1, [2,.3]).$$ ∎

Remark A.4. If $\mathcal{L}_1$ and $\mathcal{L}_2$ are two Lie algebras, and $\mathcal{G}(\mathcal{L})$ denotes the Lie algebra $\mathcal{L} \propto \mathcal{L}^*$ with the invariant metric (A.3) then

$$\mathcal{G}(\mathcal{L}_1 \oplus \mathcal{L}_2) \approx \mathcal{G}(\mathcal{L}_1) \oplus \mathcal{G}(\mathcal{L}_2), \tag{A.5}$$

$$\mathcal{G}(\mathcal{L}_1 \otimes \mathcal{L}_2) \not\approx \mathcal{G}(\mathcal{L}_1) \otimes \mathcal{G}(\mathcal{L}_2), \tag{A.6}$$

and if $h : \mathcal{L}_1 \to \mathcal{L}_2$ is a Lie homomorpphism, it does not imply a Lie homomorphism between $\mathcal{G}(\mathcal{L}_1)$ and $\mathcal{G}(\mathcal{L}_2)$. Thus, although (A.3) shows that there are as many Lie algebras with desirable metric as there are Lie algebras, the map from the latter to the former is not a functor.

Sources

§2.

The algebraic approach to the classical calculus of variations was initiated in [G –Di]. It was then generalized in [Man] by an algebraization of basic constructions from global invariant geometric calculus of variations in [Ku 1,2]. Discrete degrees of freedom (group G) were introduced into the calculus of variations in [Ku 3] (for the case $G = \mathbf{Z}^r$) and [Ku 13] (for general G). The constructions in §2 are similar in spirit to that in [Ku 9] for the even case; $\mathbf{Z}_2$-graded versions are given without proofs in [Ku 11,14]. The operator $\bar{\tau}$ (2.56) and the exact sequence (2.60) originate in geometry [Ku 2]. The observation that the (super)symmetry of the Fréchet derivative (2.66) follows from the complex (2.60) was originally made in [Man] (in the even case). The exactness proof of Theorem 2.67 is an algebraic version of the corresponding geometric proof in [Ku 2] (in the even case).

§3.

In the even case, there exist various definitions of the abstract (as opposed to naive) Hamiltonian formalism [K - M; Man; G - Do 1,2; Ku 2,9]. The $\mathbf{Z}_2$-graded version was given without proofs in [Ku 12,14]. Hamiltonian property of q-independent matrices (Theorem 3.57) dates back to [G-M-S; Man]. The criterion (3.64) for a map to be canonical is proved in [Ku 9] for the even case. Connections with Lie algebras (in the even case) were found in [G - Do 3; Ku 4, 9, 13], and in the $\mathbf{Z}_2$-graded case in [Ku 12,14]. Generalized 2-cocycles were defined in [Ku 9]. Stable Lie superalgebras were introduced in [Ku 14].

§4.

The supertrace was introduced in [Ka 1] for the case $T = \mathbf{C}$, and for general case in [Le]. Algebraic pseudo-differential operators (even) came in in [Man]; the specific form used in §4 is due to [W 1]. The symmetry of the form *tr Res* is proven in [Man]. Objects such as $d : \mathcal{O}_C \to \mathcal{O}_{\Omega^1(C)}$ were defined in [Man] (in the even case). The basic formulae (4.59) and (4.64) are derived in [Man] (also in the even case).

§5.

Classical integrable systems (for differential Lax operators) were invented by Wilson [W 1]. $\mathbf{Z}_2$-graded case is worked out in [Ku 6].

§6.

In the scalar even case ($\ell = \ell_0 = 1$), computation of variational derivatives of conservation laws and the Hamiltonian structure of Lax equations is done in [Man], and in the matrix even case ($\ell = \ell_0 > 1$) in [K - W]. Most of the methods and results in §6 are adapted to the $\mathbf{Z}_2$-graded case from [K - W]. The binomial-coefficients-free derivation of the Hamiltonian structures B_+ and B_-, and their association with rings of pseudo-differential and differential operators, respectively, is modelled on [Ku 10].

§7.

The interpretation of the KdV equation in terms of $D(K)$ was given in [Ku 7,16]. The Lie superalgebra (7.32), and the system (7.61) and its integrability, were announced during the Oberwolfach meeting "Unendlichdimensionale Lie - Algebren und Gruppen" in April, 1985. For the case $\mathcal{G} = \{0\}$, the system (7.61) was given in [Ku 5].

§9.

Bi-Hamiltonian systems appeared sporadically during the seventies (see, e.g. [Mag; G - Do 1]). The basic bi-Hamiltonian result, Theorem 9.9, is stated without proof in [G - Do 1] in the following particular case: $K = K_0 = \mathbf{C}$, $G = \{e\}$, $m = 1$. The complex structure (9.12),(9.13) was introduced into analysis of bi-superHamiltonian systems in [Ku 16]. A possibility that a super version of the Dym equation exists is mentioned in [E - O].

BIBLIOGRAPHY

Ablowitz, M. J., and Segur H.

"Solitons and the Inverse Scattering Transform", SIAM, Philadelphia (1981).

Astrakhantzev, V. V.

[1] "Symmetric Spaces of Corank 1", Mathematics of the USSR, Sbornik, 96 (1975), 135–151 (Russian); 138:1, 129–144 (English).

[2] "On Decomposability of Metrizable Lie Algebras". Funct Anal. Appl. 12:3 (1978), 64–65 (Russian); 210–212 (English).

[3] "About a Characteristic Property of Simple Lie Algebras", Funct. Anal. Appl. 19:2 (1985), 65–66 (Russian).

Cahen, M., Lemaire L., and Parker, M.

"Relèvements d'une Structure Symétrique dans des Fibrés Associés à un Espace Symètrique", Bull. Soc. Math. Belgique 24:3 (1972), 227–237.

Drinfel'd, V. G.

"Hopf Algebras and Quantum Yang–Baxter Equation", Doklady Acad. Nauk SSSR, ser. Math., v.285 No 3 (1985), 1060–1064 (in Russian).

Drinfel'd, V. G.. and Sokolov. V. V.

[1] "Equations of KdV Type and Simple Lie Algebras", Dokl. Akad. Nuak SSSR 258 (1981). 11–16 (Russian); Sov. Math. Dokl. 23 (1981), 457–462 (English).

[2] "Lie Algebras and the Korteweg de Vries Type Equations", Itogi Nauki i Tekhniki, ser. Sovremennye Problemi Mathematiki, 24 (1984), 81–180 (Russian); J. Sov. Math. 30 (1985), 1975–2036.

Erbay, S., and Oḡus, Ö.

"A Super Extension of the WKI Integrable System". J. Phys. A 18 (1985), L969–L974.

Fordy, A. P.

"Derivative Nonlinear Schrödinger Equations and Hermitian Symmetric Spaces", J. Phys. A 17 (1984), 1235–1245.

Fordy, A. P., and Kulish, P. P.
``Nonlinear Schrödinger Equations and Simple Lie Algebras'', Comm. Math. Phys. 89 (1983), 127-443.

Gel'fand, I. M., and Dorfman, I. Ya.
[1] ``Hamiltonian Operators and Related Algebraic Structures'', Funct. Anal. Appl. 13:4 (1979), 13–30 (Russian); 248–262 (English).
[2] ``The Schouten Bracket and Hamiltonian Operators'', Funct. Anal. Appl. 14 (1980), 71–74 (Russian); 223–226 (English).
[3] ``Hamiltonian Operators and Infinite-Dimensional Lie Algebras'', Funct. Anal. Appl. 15 (1981), 23–40 (Russian); 173–187 (English).

Gel'fand, I. M., and Dikii, L. A.
``Asymptotic Behaviour of the Resolvent of Sturm-Liouville Equations and the Algebra of the Korteweg- de Vries Equations'', Uspekhi Mat. Nauk 30:5 (1975), 67–100 (Russian); Russ. Math. Surv., 77–113 (English).

Gel'fand, I. M., Manin Yu. I., and Shubin, M. A.
``Poisson Brackets and the Kernel of Variational Derivatives in the Formal Calculus of Variations'', Funct. Anal. Appl. 10:4 (1976), 30–34 (Russian); 274–278 (English).

Gürses, M., and Oğuz, Ö.
``A Super AKNS Scheme'', Phys. Lett. 108A (1985), 437–440.

Jimbo, M.
``A q–Difference Analogue of $U(\mathcal{G})$ and the Yang–Baxter Equation'', Lett. Math. Phys. 10 (1985), 63-69.

Kac, V. G.
[1] ``Classification of Simple Lie Superalgebras'', Funct. Anal. Appl. 9:3 (1975), 91–92 (Russian); 263–265 (English).
[2] ``Lie Superalgebras'', Adv. Math. 26 (1976), 8-96.
[3] ``Infinite–Dimensional Lie Algebras'', Birkhäuser, Boston (1985).

Kupershmidt, B. A.
[1] ``Lagrangian Formalism in Variational Calculus'', Funct. Anal. Appl. 10:2 (1976), 77–78 (Russian); 147–149 (English).

[2] "Geometry of Jet Bundles and the Structure of Lagrangian and Hamiltonian Formalisms", in "Geometric Methods in Mathematical Physics", Lecture Notes Math. #775, Springer-Verlag (1980), 162–218.

[3] "On Algebraic Models of Dynamical Systems", Lett. Math. Phys. 6 (1982), 85–89.

[4] "On Dual Spaces of Differential Lie Algebras", Physica 7D (1983), 334–337.

[5] "A Super Korteweg-de Vries Equation: an Integrable System", Phys. Lett. 102A (1984), 213–215.

[6] "Super-Integrable Systems", Proc. Nat. Acad. Sci. USA, 81 (1984), 6562–6563.

[7] "Integrable and Superintegrable Systems, and Differential and Difference Lie Algebras and Superalgebras", in "Open Problems in the Structure Theory of Non-Linear Integrable Differential and Difference Systems", Nagoya Univ., Nagoya, Japan, pp. 14–23 (1984).

[8] "Bosons and Fermions Interacting Integrably with the Korteweg-de Vries Field", J. Phys. A 17 (1984), L869–L872.

[9] "Discrete Lax Equations and Differential-Difference Calculus", Asterisque. Paris, (1985).

[10] "Mathematics of Dispersive Water Waves", Comm. Math. Phys. 99 (1985), 51–73.

[11] "Exact Resolution of the $\mathbf{Z}_2$-Graded Euler-Lagrange Operator", Expositiones Math. 3 (1985), 375–377.

[12] "Odd and Even Poisson Brackets in Dynamical Systems". Lett. Math. Phys., 9 (1985), 323–330.

[13] "Noncommutative Calculus of Variations", pp. 297–303, in "Differential Geometry, Calculus of Variations, and their Applications", Lect. Notes Pure Appl. Math. v. 100, Marcel Dekker, New York (1985).

[14] "A Review of Superintegrable Systems". Lectures in Appl. Math. 23 Part 1 (1986), 83–120.

[15] "A Coupled Korteweg-de Vries Equation with Dispersion", J. Phys. A 18 (1985), L571–L573.

[16] ``Super Korteweg-de Vries Equations Associated to Super Extensions of the Virasoro Algebra'', Phys. Lett. 109A (1985), 417–423.

[17] ``Linearized Structures of Lagrangian, Hamiltonian, and Quasi-Hamiltonian Systems'', Phys. Lett. 114A (1986), 231–235.

Kupershmidt, B.A., and Manin, Yu. I.
``Equations of Long Waves with a Free Surface. II. Hamiltonian Structure and Higher Equations'', Funct. Anal. Appl. 12:1 (1978), 25–37 (Russian); 20–29 (English).

Kupershmidt, B.A., and Wilson, G.,
``Modifying Lax Equations and the Second Hamiltonian Structure'', Inventiones Math. 62 (1981), 403–436.

Lamb, G. L., Jr.
``Elements of Soliton Theory'', John Wiley, New York (1980).

Leites, D. A.
``Introduction to Supermanifolds'', Uspekhi Math. Nauk 35:1 (1980), 3–57 (Russian); Russ. Math. Surveys, 1–64 (English).

Magri, F.
``A Simple Model of the Integrable Hamiltonian Equation''. J. Math. Phys. 19 (1978), 1156–1162.

Manin, Yu, I.
``Algebraic Aspects of Non-Linear Differential Equations'', Itogi Nauki i Tekhniki, ser. Sovremennye Problemi Mathematiki 11 (1978), 5-152 (Russian); J. Sov. Math. 11 (1979), 1–122 (English).

Manin, Yu. I., and Radul. A. O.
``A Supersymmetric Extension of the Kadomtsev–Petviashvili Hierarchy''. Comm. Math. Phys. 98 (1985), 65–75.

McKean, H. P.
``Integrable Systems and Algebraic Curves''. Lecture Notes Math. 755. Springer–Verlag (1979), 83-200.

Medina, A.

[1] ``Structure Orthogonale sur une Algèbre de Lie et Structure de Lie–Poisson Associée'', Semin. de Geom. Diff. (1983–1984), Univers. des Sci. et Techn. du Languedoc. Montpellier.

[2] ``Groupe de Lie Munis de Pseudométriques de Riemann bi-Invariantes'', Séminaire Géom. Diff., 1981–1982. Montpellier.

[3] ``Structures de Lie–Poisson Pseudo-Riemanniennes et Structures Orthogonales C. R. Acad. Sc. Paris, t. 301, Série I (1985), 507–510.

Medina, A., et Revoy, P.

[1] ``Algèbres de Lie et Produit Scalaire Invariant''. Semin. de Geom. Diff. (1983–1984), Univers. des Sci. et Techn. du Languedoc, Montpellier.

[2] ``Charactérisation des Groupes de Lie Ayant une Pseudo–Métrique bi-Invariante'', Collection Travaux en Cours, Sém. Sud–Rhodanien, III (1984). 149–166. Hermann, Paris.

Newell, A. C.

``Solitons in Mathematics and Physics'', SIAM, Philadelphia (1985).

Novikov, S. P., ed. (Zakharov, V. E., Manakov, S. V., Novikov, S. P., and Pitaevskii, L. P.)

``Theory of Solitons. The Method of Inverse Problem'', Nauka, Moscow (1980) (in Russian).

Veselov, A. P.

``On the Hamiltonian Formalism for the Novikov–Krichever Equations of Commutativity of Two Operators''. Funct. Anal. Appl. 13 (1979), 1–7 (Russian); 1–6 (English).

Wilson, G.

[1] ``Commuting Flows and Conservations Laws for Lax Equations'', Math. Proc. Cambr. Phil. Soc. 86 (1979), 131–143.

[2] ``On Two Constructions of Conservation Laws for Lax Equations'', Quart. J. Math. Oxford (2), 32 (1981), 491–512.

[3] ``The Modified Lax and Two-Dimensional Toda Lattice Equations Associated with Simple Lie Algebras'', Ergod. Th. & Dynam. Syst. 1 (1981), 361–380.

INDEX